Construction Achievement of High-level Speciality Groups: Loose-leaf Series of Bilingual Textbooks

高水平专业群建设成果活页式系列双语教材

金工实训（活页式）

Metalworking Practical Training

主　编　邹专仁

副主编　左江林　贺敏超

主　审　邓海鹰

中国水利水电出版社

www.waterpub.com.cn

·北京·

内 容 提 要

本教材共分为电焊机临时用电、台钻床临时用电、变压器低压套管线夹M10六角螺母的加工和电气设备接地引下线的焊接四大学习情境。学习情境1和学习情境2根据工作实际场景，详细介绍了在临时用电过程中需要注意的用电安全规范、用电技术要求、确定用电方案以及实施现场用电等知识和技能要求。学习情境3主要介绍变压器低压套管线夹M10六角螺母的加工步骤和技能要求，内容以变压器低压套管线夹螺母为例，按照实际工作过程分步骤介绍金工实训中锯削技术、锉削技术、钻削技术等机加工技术要求和技能要求。学习情境4主要介绍接地装置的焊接知识和技能要求，内容以电气设备接地引下线的焊接为例，分步介绍焊接引弧技能训练—焊接运条技能训练—焊接收尾技能训练—下料—接地引下线搭接平焊等知识和技能要求。

本教材可供电力类和机电类等相关专业学生在金属加工基本技能训练时使用，也可供相关专业毕业生或社会人员在金属加工和焊接工作时参考。

图书在版编目（CIP）数据

金工实训 : 活页式 : 汉文、英文 / 邹专仁主编
. -- 北京 : 中国水利水电出版社, 2022.8
高水平专业群建设成果活页式系列双语教材
ISBN 978-7-5226-0981-2

Ⅰ. ①金… Ⅱ. ①邹… Ⅲ. ①金属加工－实习－教材
－汉、英 Ⅳ. ①TG-45

中国版本图书馆CIP数据核字(2022)第163265号

书　　名	高水平专业群建设成果活页式系列双语教材 **金工实训（活页式）** JINGONG SHIXUN (HUOYESHI)
作　　者	主　编　邹专仁 副主编　左江林　贺敏超 主　审　邓海鹰
出版发行	中国水利水电出版社 （北京市海淀区玉渊潭南路1号D座　100038） 网址：www.waterpub.com.cn E-mail：sales@mwr.gov.cn 电话：(010) 68545888（营销中心）
经　　售	北京科水图书销售有限公司 电话：(010) 68545874、63202643 全国各地新华书店和相关出版物销售网点
排　　版	中国水利水电出版社微机排版中心
印　　刷	北京市密东印刷有限公司
规　　格	184mm×260mm　16开本　20.75印张　581千字
版　　次	2022年8月第1版　2022年8月第1次印刷
印　　数	0001—1500册
定　　价	**68.00**元

Abstract

This textbook is divided into four learning scenarios: temporary electricity of electric welding machine, temporary electricity of bench drilling machine, processing the M10 hexagon nut of transformer low-voltage bushing clamp, and welding the earthing down-lead of electrical equipment. According to the actual work scenarios, learning scenarios 1 and 2 describe in detail the knowledge and skills that require attention in temporary electricity utilization, including the electricity safety code, the electricity technical requirements, the confirmation of the electricity scheme, and the on-site electricity utilization. Learning scenario 3 mainly introduces the procedures and technical requirements for processing the M10 hexagon nut of transformer low-voltage bushing. Taking the transformer low-voltage bushing clamp nut as an example, learning scenario 3 step by step introduces the machining technical and skill requirements regarding practical work procedures, including sawing, filing, drilling, etc. Learning scenario 4 mainly presents the welding knowledge of earthing devices and the technical requirements. Taking connecting electrical equipment to earthing down-lead as an example, learning scenario 4 step by step introduces the knowledge and technical requirements of welding arc striking skills training, electrode manipulation skills training, welding finishing off skills training, cutting stock, overlap flat welding of earthing down-lead, etc.

Students majoring in electrical-related and mechano-electronic-related fields can utilize this textbook in basic metalworking skills training. It can also be used as a reference by graduates in related majors or social personnel when involved in metalworking and welding work.

前　言

2019年1月，国务院印发了《国家职业教育改革实施方案》，方案指出：倡导使用新型活页式、工作手册式教材。为了适应国家职业教育改革发展需要，进一步提高学习者的理论知识和操作技能水平，我们组织学院专业带头人、骨干教师和企业专家，根据闫智勇博士关于活页式、工作手册式教材的相关研究成果，多方论证共同编制了本教材。

本教材精心筛选了4个学习情境，其特点：一是在内容上，模拟电力类和机电类专业毕业生在工作过程中常见的实际工作情境，具有很好的代表性和普遍性；二是在知识和技能教学要求上，符合电力类和机电类专业金工实训课程教学大纲要求；三是在训练流程安排上，坚持“以学习者为中心”的基本思想，有利于推动学习者借助外部资源（如教师提供的资料、互联网查询）主动学习，有效培养学习者主动分析问题和解决问题的能力。

本教材由广西水利电力职业技术学院邹专仁担任主编，广西水利电力职业技术学院左江林、贺敏超担任副主编。广西水利电力职业技术学院邓海鹰教授担任主审。参加编写的人员还有广西水利电力职业技术学院雷红梅、广西机电职业技术学院周正杰、黔西南民族职业技术学院田友飞、吉林水利电力职业技术学院郭艺和广西大通建设监理咨询管理有限公司滕贤。

本教材编写人员在编写过程中参考了闫智勇博士讲座中的“活页式和工作手册式教材编写体例”及同行们的优秀教材，得到很大启发和教益，在此表示衷心的感谢！

由于时间紧，任务重，加之编者水平有限，不足之处在所难免，敬请读者批评指正。

编者

2022年8月

Preface

In January 2019, the State Council of China issued the *National Plan for Vocational Education Reform*, which states: advocating the use of new loose-leaf and workbook-style textbooks. Therefore, we gathered professional leaders and key higher education teachers, and industry experts to jointly compile this textbook to adapt to the needs of the national vocational education reform and development and to further improve learners' theoretical knowledge and operational skills. The textbook is based on the relevant research results of Dr. Yan Zhiyong related to loose-leaf and workbook-style textbooks.

The four learning scenarios carefully selected in the textbook are characterized by three features. Firstly, in terms of content, they simulate the common actual work situations for graduates in electrical-related and mechano-electronic-related majors, with high representativeness and universality. Secondly, in terms of the teaching requirements on knowledge and skills, they meet the requirements of the teaching syllabus of metalworking practical training courses of electrical-related and mechano-electronic-related majors. Thirdly, in terms of the arrangement of the training process, they adhere to the basic "learner-centred" idea, which encourages learners to take the initiative to learn with the help of external resources (such as information provided by teachers and Internet queries), and effectively cultivates learners' ability to analyse and solve problems initiatively.

This textbook's edited-in-chief is Zou Zhuanren of the Guangxi Vocational College of Water Resources and Electric Power, with Zuo Jianglin and He Minchao from the same college, serving as associate editors. Prof. Deng Haiying of the Guangxi Vocational College of Water Resources and Electric Power was the main reviewer. Other contributors also include Lei Hongmei of the Guangxi Vocational College of Water Resources and Electric Power, Zhou Zhengjie of the Guangxi Technological College of Machinery and Electricity, Tian Youfei of the Southwest Guizhou Vocational and Technical College for Nationalities, Guo Yi of the Jilin Polytechnic of Water Resources and Electric Engineering, and Teng Xian of the Guangxi Datong Construction Consultant & Management Co. Ltd.

The authors have referenced Dr. Yan Zhiyong's lecture about loose-leaf and workbook-style textbook writing style and excellent teaching materials of other professions when compiling this textbook. We have gained great inspiration and valuable experiences from them and would like to express our sincere thanks!

Due to the time constraint and limited level of the authors' expertise, we understand that deficiencies are inevitable. We invite any comments and feedback from readers to improve our work.

Authors
August 2022

目　录

学习场

金 工 实 训

学习情境 1

电焊机临时用电

1.0 学习性工作任务单

<table>
<tr><td>学习场</td><td colspan="6">金　工　实　训</td></tr>
<tr><td>学习情境 1</td><td colspan="4">电焊机临时用电</td><td>总学时</td><td>300 分钟</td></tr>
<tr><td>典型工作过程描述</td><td colspan="6">用电安全规范—分析用电技术要求—勘察现场用电环境—确定选用方案—实施现场用电</td></tr>
<tr><td>学习目标</td><td colspan="6">1. 掌握施工现场临时用电规范。
2. 掌握施工现场电焊机用电技术要求。
3. 掌握施工现场电焊机的选用。
4. 树立安全用电职业意识，培养爱岗敬业精神。</td></tr>
<tr><td>任务描述</td><td colspan="6">某施工现场需要焊接防雷接地体（线），现按照要求接上临时电源供给电焊机使用。请根据现场需要，安全有效地完成任务，参考图 1.0.1。</td></tr>
<tr><td>学时安排</td><td>资讯
30 分钟</td><td>计划
30 分钟</td><td>决策
30 分钟</td><td>实施
160 分钟</td><td>检查
30 分钟</td><td>评价
20 分钟</td></tr>
<tr><td>对学生的要求</td><td colspan="6">课前预习实训作业单，充分利用网络查找有关实训的学习资料，实训过程中穿着实训服装，贯彻落实自己不伤害自己、自己不伤害他人、自己不被他人伤害、保护他人不被伤害的“四不伤害”和其他安全要求注意事项，严格遵守实训室的各项规章制度。</td></tr>
<tr><td>参考资料</td><td colspan="6">参考《电弧焊机通用技术条件》(GB/T 8118—2010)、《施工现场临时用电安全技术规范》(JGJ 46—2005)、《弧焊设备　第 1 部分：焊接电源》(GB/T 15579.1—2013) 和《电力安全工作规程　发电厂和变电站电气部分》(GB 26860—2011)。</td></tr>
</table>

图 1.0.1　现场临时用电

1.1　教学引导文

学习场	金工实训	学习情境 1	电焊机临时用电	参照系	《电弧焊机通用技术条件》(GB/T 8118—2010)	
普适性工作过程 典型工作过程	资讯	计划	决策	实施	检查	评价
用电安全规范	查阅《施工现场临时用电安全技术规范》(JGJ 46—2005)，获取施工现场用电规范的有关信息，便于安全有效地完成任务。	通过同学之间小组讨论，拟订用电安全规范计划，完成 1.1.1.2 用电安全规范环节计划单。	计划的可行性、经济性、可操作性和实施难度几方面综合分析，选取最佳方案。	查阅有关资料，完成 1.1.1.4 用电安全规范环节实施单。	按照 1.1.1.5 用电安全规范环节检查单的检查项目和标准，进行学生自查和教师检查。	根据评价项目配置分值，进行自评、互评和教师评价，完成 1.5 电焊机临时用电学习情境评价单。
分析用电技术要求	查阅《弧焊设备 第1部分：焊接电源》(GB/T 15579.1—2013)，获取施工现场点焊接用电技术要求的有关信息，便于安全有效地完成任务。	通过同学之间小组讨论，拟订用电安全规范计划，完成 1.1.2.2 分析用电技术要求环节计划单。	计划的可行性、经济性、可操作性和实施难度几方面综合分析，选取最佳方案。	查阅有关资料，完成 1.1.2.4 分析用电技术要求环节实施单。	按照 1.1.2.5 分析用电技术要求环节检查单的检查项目和标准，进行学生自查和教师检查。	根据评价项目配置分值，进行自评、互评和教师评价，完成 1.5 电焊机临时用电学习情境评价单。

续表

学习场	金工实训	学习情境1	电焊机临时用电	参照系	《电弧焊机通用技术条件》（GB/T 8118—2010）	
普适性工作过程 / 典型工作过程	资讯	计划	决策	实施	检查	评价
勘察现场用电环境	查阅《施工现场临时用电安全技术规范》（JGJ 46—2005），获取施工现场用电规范的有关信息，便于安全有效地完成任务。	通过同学之间小组讨论，拟订用电安全规范计划，完成1.1.3.2 勘察现场用电环境环节计划单。	计划的可行性、经济性、可操作性和实施难度几方面综合分析，选取最佳方案。	查阅有关资料，完成1.1.3.4 勘察现场用电环境环节实施单。	按照1.1.3.5 勘察现场用电环境环节检查单的检查项目和标准，进行学生自查和教师检查。	根据评价项目配置分值，进行自评、互评和教师评价，完成1.5 电焊机临时用电学习情境评价单。
确定选用方案	查阅《施工现场临时用电安全技术规范》（JGJ 46—2005），获取施工现场用电规范的有关信息，便于安全有效地完成任务。	通过同学之间小组讨论，拟订用电安全规范计划，完成1.1.4.2 确定选用方案环节计划单。	计划的可行性、经济性、可操作性和实施难度几方面综合分析，选取最佳方案。	查阅有关资料，完成1.1.4.4 确定选用方案环节实施单。	按照1.1.4.5 确定选用方案环节检查单的检查项目和标准，进行学生自查和教师检查。	根据评价项目配置分值，进行自评、互评和教师评价，完成1.5 电焊机临时用电学习情境评价单。
实施现场用电	查阅《弧焊设备　第1部分：焊接电源》（GB/T 15579.1—2013），获取施工现场点焊接用电技术要求的有关信息，便于安全有效地完成任务。	通过同学之间小组讨论，拟订用电安全规范计划，完成1.1.5.2 实施现场用电环节计划单。	计划的可行性、经济性、可操作性和实施难度几方面综合分析，选取最佳方案。	查阅有关资料，完成1.1.5.4 实施现场用电环节实施单。	按照1.1.5.5 实施现场用电环节检查单的检查项目和标准，进行学生自查和教师检查。	根据评价项目配置分值，进行自评、互评和教师评价，完成1.5 电焊机临时用电学习情境评价单。

1.1.1　用电安全规范

1.1.1.1　用电安全规范环节资讯单

<table>
<tr><td>学习场</td><td colspan="3">金　工　实　训</td></tr>
<tr><td>学习情境 1</td><td>电焊机临时用电</td><td>本环节学时</td><td>6 分钟</td></tr>
<tr><td>典型工作过程描述</td><td colspan="3">用电安全规范—分析用电技术要求—勘察现场用电环境—确定选用方案—实施现场用电</td></tr>
<tr><td>搜集资讯的方式</td><td colspan="3">教师的讲解、互联网查询或者同学间交流</td></tr>
<tr><td>资讯描述</td><td colspan="3">查看教师提供的资料，获取施工现场用电规范的有关信息，便于安全有效地完成任务，如图 1.1.1 所示。

图 1.1.1　现场勘察</td></tr>
<tr><td>对学生的要求</td><td colspan="3">课前充分预习实训作业单，充分利用网络查找有关实训的学习资料，实训过程中穿着实训服装，贯彻落实自己不伤害自己、自己不伤害他人、自己不被他人伤害、保护他人不被伤害的“四不伤害”安全要求，注意严格遵守实训室的各项规章制度。</td></tr>
<tr><td>参考资料</td><td colspan="3">参考《电力安全工作规程　发电厂和变电站电气部分》（GB 26860—2011）、《施工现场临时用电安全技术规范》（JGJ 46—2005）和《弧焊设备　第 1 部分：焊接电源》（GB/T 15579.1—2013）。</td></tr>
</table>

1.1.1.2 用电安全规范环节计划单

<table>
<tr><td colspan="2">学习场</td><td colspan="5">金 工 实 训</td></tr>
<tr><td colspan="2">学习情境1</td><td colspan="3">电焊机临时用电</td><td>本环节学时</td><td>6分钟</td></tr>
<tr><td colspan="2">典型工作过程描述</td><td colspan="5">用电安全规范—分析用电技术要求—勘察现场用电环境—确定选用方案—实施现场用电</td></tr>
<tr><td colspan="2">计划制订的方式</td><td colspan="5">同学之间小组讨论</td></tr>
<tr><td>序号</td><td colspan="2">工 作 步 骤</td><td colspan="4">注 意 事 项</td></tr>
<tr><td>1</td><td colspan="2">开班前会进行用电安全教育</td><td colspan="4">查看工作班成员的精神面貌是否良好，了解大家对用电安全规范的掌握情况，培养作业人员的职业安全意识和团队协作精神，树立爱岗敬岗的职业价值观。</td></tr>
<tr><td>2</td><td colspan="2">观看现场触电救护视频</td><td colspan="4"></td></tr>
<tr><td>3</td><td colspan="2">学习施工现场用电安全知识、有关国家行业标准和文件精神</td><td colspan="4"></td></tr>
<tr><td>4</td><td colspan="2">讨论学习的收获</td><td colspan="4"></td></tr>
<tr><td>5</td><td colspan="2">处理分析触电或者火灾等其他应急事故</td><td colspan="4"></td></tr>
<tr><td colspan="2" rowspan="3">计划评价</td><td>班级</td><td></td><td>第（ ）组</td><td>组长签字</td><td></td></tr>
<tr><td>教师签字</td><td></td><td>日期</td><td colspan="2"></td></tr>
<tr><td colspan="5">评语：</td></tr>
</table>

1.1.1.3　用电安全规范环节决策单

学习场	金　工　实　训				
学习情境 1	电焊机临时用电			本环节学时	6 分钟
典型工作过程描述	用电安全规范—分析用电技术要求—勘察现场用电环境—确定选用方案—实施现场用电				
计　划　对　比					
序号	计划的可行性	计划的经济性	计划的可操作性	计划的实施难度	综合评价
方案 1					
方案 2					
方案 3					
决策评价	班级		第（　　）组	组长签字	
	教师签字		日期		
	评语：				

1.1.1.4　用电安全规范环节实施单

<table>
<tr><td colspan="2">学习场</td><td colspan="4">金　工　实　训</td></tr>
<tr><td colspan="2">学习情境1</td><td colspan="2">电焊机临时用电</td><td>本环节学时</td><td>30分钟</td></tr>
<tr><td colspan="2">典型工作过程描述</td><td colspan="4">用电安全规范—分析用电技术要求—勘察现场用电环境—确定选用方案—实施现场用电</td></tr>
<tr><td>序号</td><td>实施步骤</td><td colspan="3">工作内容及方法</td><td>注意事项</td></tr>
<tr><td>1</td><td>开班前会</td><td colspan="3">开班前会进行用电安全教育，查看工作班成员的精神面貌是否良好，了解大家对用电安全规范的掌握情况，培养作业人员的职业安全意识和团队协作精神，树立爱岗敬岗的职业价值观。</td><td></td></tr>
<tr><td>2</td><td>观看视频</td><td colspan="3">观看现场触电救护视频。</td><td>触电急救技能</td></tr>
<tr><td>3</td><td>学习文件标准</td><td colspan="3">学习《电力安全工作规程　发电厂和变电站电气部分》(GB 26860—2011)和《施工现场临时用电安全技术规范》(JGJ 46—2005)。查找并学习临时用电配电箱系统图等。</td><td></td></tr>
<tr><td>4</td><td>谈学习收获</td><td colspan="3"></td><td></td></tr>
<tr><td>5</td><td>应急事故处理分析</td><td colspan="3"></td><td></td></tr>
<tr><td colspan="6">实施说明：</td></tr>
<tr><td rowspan="3" colspan="2">实施评价</td><td>班级</td><td></td><td>第（　　）组</td><td>组长签字　</td></tr>
<tr><td>教师签字</td><td></td><td>日期</td><td></td></tr>
<tr><td colspan="4">评语：</td></tr>
</table>

1.1.1.5 用电安全规范环节检查单

<table>
<tr><td colspan="2">学习场</td><td colspan="3">金 工 实 训</td></tr>
<tr><td colspan="2">学习情境 1</td><td>电焊机临时用电</td><td>本环节学时</td><td>6 分钟</td></tr>
<tr><td colspan="2">典型工作过程描述</td><td colspan="3">用电安全规范—分析用电技术要求—勘察现场用电环境—确定选用方案—实施现场用电</td></tr>
<tr><td>序号</td><td>检查项目</td><td>检 查 标 准</td><td>学生自查</td><td>教师检查</td></tr>
<tr><td>1</td><td>用电安全资讯环节</td><td>获得施工现场用电安全规范的有关信息的来源，是否是电力行业用电安全规范标准。</td><td></td><td></td></tr>
<tr><td>2</td><td>用电安全计划环节</td><td>根据施工现场已有条件和具体情况制定用电安全规范步骤的各个方案。</td><td></td><td></td></tr>
<tr><td>3</td><td>用电安全决策环节</td><td>对各种用电安全规范的方案衡量对比评分，选择最佳方案。</td><td></td><td></td></tr>
<tr><td>4</td><td>用电安全实施环节</td><td>用电安全规范过程中各个环节是否按照国家行业标准执行，是否贯彻落实“四不伤害”。</td><td></td><td></td></tr>
<tr><td>5</td><td>用电安全检查环节</td><td>各个环节逐一检查，是否有爱岗敬业的主题思想。</td><td></td><td></td></tr>
<tr><td rowspan="3">检查评价</td><td>班级</td><td></td><td>第（ ）组</td><td>组长签字</td><td></td></tr>
<tr><td>教师签字</td><td></td><td>日期</td><td colspan="2"></td></tr>
<tr><td colspan="5">评语：</td></tr>
</table>

1.1.2　分析用电技术要求

1.1.2.1　分析用电技术要求环节资讯单

<table>
<tr><td>学习场</td><td colspan="3">金　工　实　训</td></tr>
<tr><td>学习情境 1</td><td>电焊机临时用电</td><td>本环节学时</td><td>6 分钟</td></tr>
<tr><td>典型工作过程描述</td><td colspan="3">用电安全规范—分析用电技术要求—勘察现场用电环境—确定选用方案—实施现场用电</td></tr>
<tr><td>搜集资讯的方式</td><td colspan="3">教师的讲解、互联网查询或者同学间交流</td></tr>
<tr><td>资讯描述</td><td colspan="3">查看教师提供的资料，获取施工现场电焊机用电安全规范的有关信息，便于安全有效地完成任务，如图 1.1.2 所示。

图 1.1.2　查找资料</td></tr>
<tr><td>对学生的要求</td><td colspan="3">课前预习实训作业单，利用网络查找有关实训的学习资料，实训过程中穿着实训服装，贯彻落实自己不伤害自己、自己不伤害他人、自己不被他人伤害、保护他人不被伤害的“四不伤害”安全要求，注意严格遵守实训室的各项规章制度。</td></tr>
<tr><td>参考资料</td><td colspan="3">参考《电力安全工作规程　发电厂和变电站电气部分》（GB 26860—2011）和《施工现场临时用电安全技术规范》（JGJ 46—2005）。</td></tr>
</table>

1.1.2.2 分析用电技术要求环节计划单

<table>
<tr><td>学习场</td><td colspan="5">金 工 实 训</td></tr>
<tr><td>学习情境1</td><td colspan="3">电焊机临时用电</td><td>本环节学时</td><td>6分钟</td></tr>
<tr><td>典型工作过程描述</td><td colspan="5">用电安全规范—分析用电技术要求—勘察现场用电环境—确定选用方案—实施现场用电</td></tr>
<tr><td>计划制订的方式</td><td colspan="5">同学之间小组讨论</td></tr>
<tr><td>序号</td><td>工作步骤</td><td colspan="4">注意事项</td></tr>
<tr><td>1</td><td>查找现场提供的电焊机参数</td><td colspan="4">电焊机使用的是单相交流电还是三相交流电，电流的调节范围等。</td></tr>
<tr><td>2</td><td>分析电焊机正常电流是否满足焊件焊接的技术要求</td><td colspan="4"></td></tr>
<tr><td>3</td><td>检测电焊机的绝缘性能</td><td colspan="4">1. 对绝缘摇表进行自检，包括短路试验和开路试验。
2. 摇表保持转速在120r/min。
3. 注意摇表转速恒定时再将“L”端接入。
4. 待摇表指针稳定后再读数。
5. 测完后要先将“L”端拆出，再停止摇动。</td></tr>
<tr><td>4</td><td>分析配电箱空气开关工作电流是否与电焊机工作电流匹配</td><td colspan="4"></td></tr>
<tr><td>5</td><td>分析电焊机电源线是否符合规范标准</td><td colspan="4"></td></tr>
<tr><td>6</td><td>分析电焊机接地线的安装是否符合要求</td><td colspan="4"></td></tr>
<tr><td rowspan="3">计划评价</td><td>班级</td><td></td><td>第（ ）组</td><td>组长签字</td><td></td></tr>
<tr><td>教师签字</td><td></td><td>日期</td><td colspan="2"></td></tr>
<tr><td colspan="5">评语：</td></tr>
</table>

1.1.2.3 分析用电技术要求环节决策单

<table>
<tr><td>学习场</td><td colspan="5">金 工 实 训</td></tr>
<tr><td>学习情境 1</td><td colspan="3">电焊机临时用电</td><td>本环节学时</td><td>6 分钟</td></tr>
<tr><td>典型工作过程描述</td><td colspan="5">用电安全规范—分析用电技术要求—勘察现场用电环境—确定选用方案—实施现场用电</td></tr>
<tr><td colspan="6">计 划 对 比</td></tr>
<tr><td>序号</td><td>计划的可行性</td><td>计划的经济性</td><td>计划的可操作性</td><td>计划的实施难度</td><td>综合评价</td></tr>
<tr><td>方案 1</td><td></td><td></td><td></td><td></td><td></td></tr>
<tr><td>方案 2</td><td></td><td></td><td></td><td></td><td></td></tr>
<tr><td>方案 3</td><td></td><td></td><td></td><td></td><td></td></tr>
<tr><td rowspan="3">决策评价</td><td>班级</td><td></td><td>第（ ）组</td><td>组长签字</td><td></td></tr>
<tr><td>教师签字</td><td></td><td>日期</td><td colspan="2"></td></tr>
<tr><td colspan="5">评语：</td></tr>
</table>

1.1.2.4 分析用电技术要求环节实施单

<table>
<tr><td>学习场</td><td colspan="4">金 工 实 训</td></tr>
<tr><td>学习情境 1</td><td colspan="2">电焊机临时用电</td><td>本环节学时</td><td>30 分钟</td></tr>
<tr><td>典型工作过程描述</td><td colspan="4">用电安全规范—分析用电技术要求—勘察现场用电环境—确定选用方案—实施现场用电</td></tr>
<tr><td>序号</td><td>实施步骤</td><td colspan="2">工作内容及方法</td><td>注意事项</td></tr>
<tr><td>1</td><td>查阅现场的电焊机参数</td><td colspan="2"></td><td></td></tr>
<tr><td>2</td><td>分析电焊机正常电流是否满足焊件焊接的技术要求</td><td colspan="2"></td><td></td></tr>
<tr><td>3</td><td>检测电焊机的绝缘性能</td><td colspan="2">1. 对摇表进行短路试验和开路试验自检。
2. 测试一次侧对地绝缘电阻值，把“E”端接在焊机外壳，摇表保持转速在 120r/min；再将“L”端接到一次侧线圈，指针稳定后再读数，测完后要先将“L”端拆出，再停止摇动。
3. 用同样方法测试二次侧线圈对地和一次侧和二次侧线圈之间的绝缘电阻值。测试值是否符合国家电力行业的标准。</td><td></td></tr>
<tr><td>4</td><td>分析配电箱空气开关工作电流是否与电焊机工作电流匹配</td><td colspan="2"></td><td></td></tr>
<tr><td>5</td><td>分析电焊机电源线是否符合规范标准</td><td colspan="2"></td><td></td></tr>
<tr><td>6</td><td>分析电焊机接地线的安装是否符合要求</td><td colspan="2"></td><td></td></tr>
<tr><td colspan="5">实施说明</td></tr>
<tr><td rowspan="3">实施评价</td><td>班级</td><td></td><td>第（ ）组</td><td>组长签字</td></tr>
<tr><td>教师签字</td><td></td><td>日期</td><td></td></tr>
<tr><td colspan="4">评语：</td></tr>
</table>

1.1.2.5 分析用电技术要求环节检查单

学习场	金 工 实 训				
学习情境 1	电焊机临时用电			本环节学时	6 分钟
典型工作过程描述	用电安全规范—分析用电技术要求—勘察现场用电环境—确定选用方案—实施现场用电				
序号	检查项目	检 查 标 准		学生自查	教师检查
1	资讯环节	获得施工现场满足电焊机用电安全的有关信息的来源是否符合国家电力行业标准。			
2	计划环节	制定施工现场满足电焊机用电的各个方案。			
3	决策环节	对各种施工现场满足电焊机用电的方案衡量对比评分，选取最佳方案。			
4	实施环节	施工现场电焊机用电过程中各个环节按照标准执行，贯彻落实“四不伤害”。			
5	检查环节	各个环节逐一检查，是否有爱岗敬业的主题思想。			
检查评价	班级		第（ ）组	组长签字	
	教师签字		日期		
	评语：				

1.1.3　勘察现场用电环境

1.1.3.1　勘察现场用电环境环节资讯单

<table>
<tr><td>学习场</td><td colspan="3">金　工　实　训</td></tr>
<tr><td>学习情境 1</td><td>电焊机临时用电</td><td>本环节学时</td><td>6 分钟</td></tr>
<tr><td>典型工作过程描述</td><td colspan="3">用电安全规范—分析用电技术要求—勘察现场用电环境—确定选用方案—实施现场用电</td></tr>
<tr><td>搜集资讯的方式</td><td colspan="3">教师的讲解、互联网查询或者同学间交流</td></tr>
<tr><td>资讯描述</td><td colspan="3">查看教师提供的资料，获取勘察施工现场电焊机用电环境的有关信息，便于安全有效地完成任务，如图 1.1.3 所示。

图 1.1.3　勘察现场用电环境</td></tr>
<tr><td>对学生的要求</td><td colspan="3">课前充分预习实训作业单，充分利用网络查找有关实训的学习资料，实训过程中穿着实训服装，贯彻落实自己不伤害自己、自己不伤害他人、自己不被他人伤害、保护他人不被伤害的“四不伤害”安全要求，注意严格遵守实训室的各项规章制度。</td></tr>
<tr><td>参考资料</td><td colspan="3">参考《电力安全工作规程　发电厂和变电站电气部分》（GB 26860—2011）和《施工现场临时用电安全技术规范》（JGJ 46—2005）。</td></tr>
</table>

1.1.3.2 勘察现场用电环境环节计划单

<table>
<tr><td colspan="2">学习场</td><td colspan="5">金 工 实 训</td></tr>
<tr><td colspan="2">学习情境 1</td><td colspan="3">电焊机临时用电</td><td>本环节学时</td><td>6 分钟</td></tr>
<tr><td colspan="2">典型工作过程描述</td><td colspan="5">用电安全规范—分析用电技术要求—勘察现场用电环境—确定选用方案—实施现场用电</td></tr>
<tr><td colspan="2">计划制订的方式</td><td colspan="5">同学之间小组讨论</td></tr>
<tr><td>序号</td><td>工 作 步 骤</td><td colspan="5">注 意 事 项</td></tr>
<tr><td>1</td><td>进入施工现场前准备工作</td><td colspan="5">检查着装是否整齐，是否戴好安全帽和其他个人防护用品等。</td></tr>
<tr><td>2</td><td>进入施工现场</td><td colspan="5"></td></tr>
<tr><td>3</td><td>现场勘察工作</td><td colspan="5">查看焊接作业点是否配置有足够数量的消防设备；查看焊接作业点与氧气瓶、电石桶和乙炔发生器在存放、使用过程中是否符合规定的安全距离；查看现有电源距离焊接作业点是否符合要求；查看电源开关是否配有漏电装置；查看焊接工作地面是否符合要求；查看现场其他防护措施是否到位。</td></tr>
<tr><td>4</td><td>离开施工现场</td><td colspan="5"></td></tr>
<tr><td colspan="2" rowspan="3">计划评价</td><td>班级</td><td></td><td>第（ ）组</td><td>组长签字</td><td></td></tr>
<tr><td>教师签字</td><td></td><td>日期</td><td colspan="2"></td></tr>
<tr><td colspan="5">评语：</td></tr>
</table>

1.1.3.3　勘察现场用电环境环节决策单

<table>
<tr><td>学习场</td><td colspan="6">金　工　实　训</td></tr>
<tr><td>学习情境 1</td><td colspan="3">电焊机临时用电</td><td colspan="2">本环节学时</td><td>6 分钟</td></tr>
<tr><td>典型工作过程描述</td><td colspan="6">用电安全规范—分析用电技术要求—勘察现场用电环境—确定选用方案—实施现场用电</td></tr>
<tr><td colspan="7">计　划　对　比</td></tr>
<tr><td>序号</td><td colspan="2">计划的可行性</td><td>计划的经济性</td><td>计划的可操作性</td><td>计划的实施难度</td><td>综合评价</td></tr>
<tr><td>方案 1</td><td colspan="2"></td><td></td><td></td><td></td><td></td></tr>
<tr><td>方案 2</td><td colspan="2"></td><td></td><td></td><td></td><td></td></tr>
<tr><td>方案 3</td><td colspan="2"></td><td></td><td></td><td></td><td></td></tr>
<tr><td rowspan="3">决策评价</td><td>班级</td><td></td><td>第（　　）组</td><td>组长签字</td><td colspan="2"></td></tr>
<tr><td>教师签字</td><td></td><td>日期</td><td colspan="3"></td></tr>
<tr><td colspan="6">评语：</td></tr>
</table>

1.1.3.4 勘察现场用电环境环节实施单

<table>
<tr><td>学习场</td><td colspan="5">金 工 实 训</td></tr>
<tr><td>学习情境 1</td><td colspan="3">电焊机临时用电</td><td>本环节学时</td><td>30 分钟</td></tr>
<tr><td>典型工作过程描述</td><td colspan="5">用电安全规范—分析用电技术要求—勘察现场用电环境—确定选用方案—实施现场用电</td></tr>
<tr><td>序号</td><td>实施步骤</td><td colspan="2">工作内容及方法</td><td colspan="2">注意事项</td></tr>
<tr><td>1</td><td>进入施工现场前准备工作</td><td colspan="2">检查着装是否整齐，是否戴好安全帽和其他个人防护用品等。</td><td colspan="2"></td></tr>
<tr><td>2</td><td>进入施工现场</td><td colspan="2"></td><td colspan="2"></td></tr>
<tr><td>3</td><td>现场勘察工作</td><td colspan="2">查看焊接作业点是否配置有足够数量的消防设备；查看焊接作业点与氧气瓶、电石桶和乙炔发生器在存放、使用过程中是否符合规定的安全距离；查看现有电源距离焊接作业点是否符合要求；查看电源开关是否配有漏电装置；查看焊接工作地面是否符合要求；查看现场其他防护措施是否到位。</td><td colspan="2"></td></tr>
<tr><td>4</td><td>离开施工现场</td><td colspan="2"></td><td colspan="2"></td></tr>
<tr><td colspan="6">实施说明：</td></tr>
</table>

<table>
<tr><td rowspan="3">实施评价</td><td>班级</td><td></td><td>第（　　）组</td><td>组长签字</td><td></td></tr>
<tr><td>教师签字</td><td></td><td>日期</td><td colspan="2"></td></tr>
<tr><td colspan="5">评语：</td></tr>
</table>

1.1.3.5　勘察现场用电环境环节检查单

<table>
<tr><td colspan="2">学习场</td><td colspan="4">金　工　实　训</td></tr>
<tr><td colspan="2">学习情境 1</td><td colspan="2">电焊机临时用电</td><td>本环节学时</td><td>6 分钟</td></tr>
<tr><td colspan="2">典型工作过程描述</td><td colspan="4">用电安全规范—分析用电技术要求—勘察现场用电环境—确定选用方案—实施现场用电</td></tr>
<tr><td>序号</td><td>检查项目</td><td colspan="2">检 查 标 准</td><td>学生自查</td><td>教师检查</td></tr>
<tr><td>1</td><td>资讯环节</td><td colspan="2">获得勘察施工现场电焊机用电环境的有关信息的来源。</td><td></td><td></td></tr>
<tr><td>2</td><td>计划环节</td><td colspan="2">制定勘察施工现场电焊机用电环境步骤的各个方案。</td><td></td><td></td></tr>
<tr><td>3</td><td>决策环节</td><td colspan="2">对各种勘察施工现场电焊机用电环境的方案衡量对比评分，获取最佳方案。</td><td></td><td></td></tr>
<tr><td>4</td><td>实施环节</td><td colspan="2">勘察施工现场电焊机用电环境过程中各个环节按照标准执行，贯彻落实“四不伤害”。</td><td></td><td></td></tr>
<tr><td>5</td><td>检查环节</td><td colspan="2">各个环节逐一检查，是否有爱岗敬业的主题思想。</td><td></td><td></td></tr>
<tr><td colspan="2" rowspan="3">检查评价</td><td>班级</td><td></td><td>第（　　）组</td><td>组长签字　</td></tr>
<tr><td>教师签字</td><td></td><td>日期</td><td></td></tr>
<tr><td colspan="4">评语：</td></tr>
</table>

1.1.4 确定选用方案

1.1.4.1 确定选用方案环节资讯单

<table>
<tr><td>学习场</td><td colspan="3">金 工 实 训</td></tr>
<tr><td>学习情境 1</td><td>电焊机临时用电</td><td>本环节学时</td><td>6 分钟</td></tr>
<tr><td>典型工作过程描述</td><td colspan="3">用电安全规范—分析用电技术要求—勘察现场用电环境—确定选用方案—实施现场用电</td></tr>
<tr><td>搜集资讯的方式</td><td colspan="3">教师的讲解、互联网查询或者同学间交流</td></tr>
<tr><td>资讯描述</td><td colspan="3">查看教师提供的资料，获取施工现场确定电焊机选用方案的有关信息，便于安全有效地完成任务，如图 1.1.4 所示。

图 1.1.4 确定方案</td></tr>
<tr><td>对学生的要求</td><td colspan="3">课前充分预习实训作业单，充分利用网络查找有关实训的学习资料，实训过程中穿着实训服装，贯彻落实自己不伤害自己、自己不伤害他人、自己不被他人伤害、保护他人不被伤害的“四不伤害”安全要求，注意严格遵守实训室的各项规章制度。</td></tr>
<tr><td>参考资料</td><td colspan="3">参考《电力安全工作规程 发电厂和变电站电气部分》（GB 26860—2011）和《施工现场临时用电安全技术规范》（JGJ 46—2005）。</td></tr>
</table>

1.1.4.2　确定选用方案环节计划单

<table>
<tr><td>学习场</td><td colspan="5">金　工　实　训</td></tr>
<tr><td>学习情境 1</td><td colspan="3">电焊机临时用电</td><td>本环节学时</td><td>6 分钟</td></tr>
<tr><td>典型工作过程描述</td><td colspan="5">用电安全规范—分析用电技术要求—勘察现场用电环境—确定选用方案—实施现场用电</td></tr>
<tr><td>计划制订的方式</td><td colspan="5">同学之间小组讨论</td></tr>
<tr><td>序号</td><td>工 作 步 骤</td><td colspan="4">注 意 事 项</td></tr>
<tr><td>1</td><td>查询有关国家电力行业的标准</td><td colspan="4"></td></tr>
<tr><td>2</td><td>编写方案</td><td colspan="4"></td></tr>
<tr><td>3</td><td>提交方案</td><td colspan="4"></td></tr>
<tr><td>4</td><td>小组讨论</td><td colspan="4"></td></tr>
<tr><td>5</td><td>确定方案</td><td colspan="4"></td></tr>
<tr><td rowspan="3">计划评价</td><td>班级</td><td></td><td>第（　　）组</td><td>组长签字</td><td></td></tr>
<tr><td>教师签字</td><td></td><td>日期</td><td colspan="2"></td></tr>
<tr><td colspan="5">评语：</td></tr>
</table>

1.1.4.3 确定选用方案环节决策单

<table>
<tr><td>学习场</td><td colspan="6">金　工　实　训</td></tr>
<tr><td>学习情境1</td><td colspan="3">电焊机临时用电</td><td colspan="2">本环节学时</td><td>6分钟</td></tr>
<tr><td>典型工作过程描述</td><td colspan="6">用电安全规范—分析用电技术要求—勘察现场用电环境—确定选用方案—实施现场用电</td></tr>
<tr><td colspan="7">计　划　对　比</td></tr>
<tr><td colspan="2">序号</td><td>计划的可行性</td><td>计划的经济性</td><td>计划的可操作性</td><td>计划的实施难度</td><td>综合评价</td></tr>
<tr><td colspan="2">方案1</td><td></td><td></td><td></td><td></td><td></td></tr>
<tr><td colspan="2">方案2</td><td></td><td></td><td></td><td></td><td></td></tr>
<tr><td colspan="2">方案3</td><td></td><td></td><td></td><td></td><td></td></tr>
<tr><td rowspan="3">决策评价</td><td>班级</td><td></td><td>第（　　）组</td><td>组长签字</td><td colspan="2"></td></tr>
<tr><td>教师签字</td><td></td><td>日期</td><td colspan="3"></td></tr>
<tr><td colspan="6">评语：</td></tr>
</table>

1.1.4.4　确定选用方案环节实施单

<table>
<tr><td>学习场</td><td colspan="5">金　工　实　训</td></tr>
<tr><td>学习情境 1</td><td colspan="3">电焊机临时用电</td><td>本环节学时</td><td>30 分钟</td></tr>
<tr><td>典型工作过程描述</td><td colspan="5">用电安全规范—分析用电技术要求—勘察现场用电环境—确定选用方案—实施现场用电</td></tr>
<tr><td>序号</td><td>实施步骤</td><td colspan="2">工作内容及方法</td><td colspan="2">注意事项</td></tr>
<tr><td>1</td><td>查询有关标准</td><td colspan="2"></td><td colspan="2"></td></tr>
<tr><td>2</td><td>编写方案</td><td colspan="2"></td><td colspan="2"></td></tr>
<tr><td>3</td><td>提交方案</td><td colspan="2"></td><td colspan="2"></td></tr>
<tr><td>4</td><td>小组讨论</td><td colspan="2"></td><td colspan="2"></td></tr>
<tr><td>5</td><td>确定方案</td><td colspan="2"></td><td colspan="2"></td></tr>
<tr><td colspan="6">实施说明：</td></tr>
<tr><td rowspan="3">实施评价</td><td>班级</td><td></td><td>第（　　）组</td><td>组长签字</td><td></td></tr>
<tr><td>教师签字</td><td></td><td>日期</td><td colspan="2"></td></tr>
<tr><td colspan="5">评语：</td></tr>
</table>

1.1.4.5 确定选用方案环节检查单

学习场	金工实训				
学习情境 1	电焊机临时用电			本环节学时	6 分钟
典型工作过程描述	用电安全规范—分析用电技术要求—勘察现场用电环境—确定选用方案—实施现场用电				
序号	检查项目	检查标准		学生自查	教师检查
1	确定选用方案资讯环节	获得施工现场电焊机选用的有关信息的来源。			
2	确定选用方案计划环节	制定施工现场电焊机选用步骤的各个方案。			
3	确定选用方案决策环节	对各种施工现场电焊机选用的方案衡量对比评分，获取最佳方案。			
4	确定选用方案实施环节	施工现场电焊机选用过程中各个环节按照标准执行，贯彻落实“四不伤害”。			
5	确定选用方案检查环节	各个环节逐一检查，是否有爱岗敬业的主题思想。			
检查评价	班级		第（　　）组	组长签字	
	教师签字		日期		
	评语：				

1.1.5　实施现场用电

1.1.5.1　实施现场用电环节资讯单

<table>
<tr><td>学习场</td><td colspan="3">金　工　实　训</td></tr>
<tr><td>学习情境 1</td><td>电焊机临时用电</td><td>本环节学时</td><td>6 分钟</td></tr>
<tr><td>典型工作过程描述</td><td colspan="3">用电安全规范—分析用电技术要求—勘察现场用电环境—确定选用方案—实施现场用电</td></tr>
<tr><td>搜集资讯的方式</td><td colspan="3">教师的讲解、互联网查询或者同学间交流</td></tr>
<tr><td>资讯描述</td><td colspan="3">查看教师提供的资料，获取现场电焊机用电的有关信息，便于安全有效地完成任务，如图 1.1.5 所示。

图 1.1.5　现场用电</td></tr>
<tr><td>对学生的要求</td><td colspan="3">课前充分预习实训作业单，充分利用网络查找有关实训的学习资料，实训过程中穿着实训服装，贯彻落实自己不伤害自己、自己不伤害他人、自己不被他人伤害、保护他人不被伤害的“四不伤害”安全要求，注意严格遵守实训室的各项规章制度。</td></tr>
<tr><td>参考资料</td><td colspan="3">参考《电力安全工作规程　发电厂和变电站电气部分》（GB 26860—2011）和《施工现场临时用电安全技术规范》（JGJ 46—2005）。</td></tr>
</table>

1.1.5.2　实施现场用电环节计划单

<table>
<tr><td colspan="2">学习场</td><td colspan="4">金　工　实　训</td></tr>
<tr><td colspan="2">学习情境 1</td><td colspan="2">电焊机临时用电</td><td>本环节学时</td><td>6 分钟</td></tr>
<tr><td colspan="2">典型工作过程描述</td><td colspan="4">用电安全规范—分析用电技术要求—勘察现场用电环境—确定选用方案—实施现场用电</td></tr>
<tr><td colspan="2">计划制订的方式</td><td colspan="4">同学之间小组讨论</td></tr>
<tr><td>序号</td><td>工 作 步 骤</td><td colspan="3">工 作 内 容</td><td>注意事项</td></tr>
<tr><td>1</td><td>作业前准备，落实现场安全措施</td><td colspan="3">检查工作人员的精神状态；办理工作许可手续；工作负责人进行现场安全交底，向工作人员指明工作任务、工作范围、带电部位及安全措施情况。</td><td></td></tr>
<tr><td>2</td><td>停电</td><td colspan="3"></td><td></td></tr>
<tr><td>3</td><td>验电</td><td colspan="3"></td><td></td></tr>
<tr><td>4</td><td>安装设备、接线</td><td colspan="3"></td><td></td></tr>
<tr><td>5</td><td>送电</td><td colspan="3"></td><td></td></tr>
<tr><td>6</td><td>焊接</td><td colspan="3"></td><td></td></tr>
<tr><td colspan="2" rowspan="3">计划评价</td><td>班级</td><td></td><td>第（　　）组</td><td>组长签字</td></tr>
<tr><td>教师签字</td><td></td><td>日期</td><td></td></tr>
<tr><td colspan="4">评语：</td></tr>
</table>

1.1.5.3 实施现场用电环节决策单

<table>
<tr><td>学习场</td><td colspan="5">金 工 实 训</td></tr>
<tr><td>学习情境 1</td><td colspan="3">电焊机临时用电</td><td>本环节学时</td><td>6 分钟</td></tr>
<tr><td>典型工作过程描述</td><td colspan="5">用电安全规范—分析用电技术要求—勘察现场用电环境—确定选用方案—实施现场用电</td></tr>
<tr><td colspan="6">计 划 对 比</td></tr>
<tr><td>序号</td><td>计划的可行性</td><td>计划的经济性</td><td>计划的可操作性</td><td>计划的实施难度</td><td>综合评价</td></tr>
<tr><td>方案 1</td><td></td><td></td><td></td><td></td><td></td></tr>
<tr><td>方案 2</td><td></td><td></td><td></td><td></td><td></td></tr>
<tr><td>方案 3</td><td></td><td></td><td></td><td></td><td></td></tr>
<tr><td rowspan="3">决策评价</td><td>班级</td><td></td><td>第（ ）组</td><td>组长签字</td><td></td></tr>
<tr><td>教师签字</td><td></td><td>日期</td><td colspan="2"></td></tr>
<tr><td colspan="5">评语：</td></tr>
</table>

1.1.5.4 实施现场用电环节实施单

学习场	金 工 实 训		
学习情境 1	电焊机临时用电	本环节学时	40 分钟
典型工作过程描述	用电安全规范—分析用电技术要求—勘察现场用电环境—确定选用方案—实施现场用电		

序号	实施步骤	工作内容及方法	注意事项
1	作业前准备，落实现场安全措施	检查工作人员的精神状态；办理工作许可手续；工作负责人进行现场安全交底，向工作人员指明工作任务、工作范围、带电部位及安全措施情况。	预防触电事故发生。
2	停电	执行一人操作一人监护的制度。操作者戴上安全帽、戴好绝缘手套，把电源开关拉下。	预防触电事故发生。
3	验电		
4	安装设备、接线		
5	送电		
6	焊接		
实施说明：			

实施评价	班级		第（ ）组	组长签字	
	教师签字		日期		
	评语：				

不经历风雨，怎么见彩虹。苦练千百次，操作如有神。

1.1.5.5 实施现场用电环节检查单

<table>
<tr><td>学习场</td><td colspan="5">金 工 实 训</td></tr>
<tr><td>学习情境 1</td><td colspan="2">电焊机临时用电</td><td>本环节学时</td><td colspan="2">6 分钟</td></tr>
<tr><td>典型工作过程描述</td><td colspan="5">用电安全规范—分析用电技术要求—勘察现场用电环境—确定选用方案—实施现场用电</td></tr>
<tr><td>序号</td><td>检查项目</td><td colspan="2">检 查 标 准</td><td>学生自查</td><td>教师检查</td></tr>
<tr><td>1</td><td>资讯环节</td><td colspan="2">获得施工现场电焊机用电的有关信息的来源。</td><td></td><td></td></tr>
<tr><td>2</td><td>计划环节</td><td colspan="2">制定施工现场电焊机用电步骤的各个方案。</td><td></td><td></td></tr>
<tr><td>3</td><td>决策环节</td><td colspan="2">对各种施工现场电焊机用电的方案衡量对比评分，获取最佳方案。</td><td></td><td></td></tr>
<tr><td>4</td><td>实施环节</td><td colspan="2">施工现场电焊机用电中各个环节按照标准执行，贯彻落实“四不伤害”。</td><td></td><td></td></tr>
<tr><td>5</td><td>检查环节</td><td colspan="2">各个环节逐一检查，是否有爱岗敬业的思想主题。</td><td></td><td></td></tr>
<tr><td rowspan="3">检查评价</td><td>班级</td><td></td><td>第（ ）组</td><td>组长签字</td><td></td></tr>
<tr><td>教师签字</td><td></td><td>日期</td><td colspan="2"></td></tr>
<tr><td colspan="5">评语：</td></tr>
</table>

1.2 电焊机临时用电学习情境材料工具清单

学习场	金工实训		
学习情境 1	电焊机临时用电	本环节学时	300 分钟
典型工作过程描述	用电安全规范—分析用电技术要求—勘察现场用电环境—确定选用方案—实施现场用电		

序号	名称	作用	数量	型号	使用者

班级		第（ ）组	组长签字	
教师签字		日期		

1.3　电焊机临时用电学习情境教学反馈单

<table>
<tr><td>学习场</td><td colspan="4">金　工　实　训</td></tr>
<tr><td>学习情境 1</td><td colspan="2">电焊机临时用电</td><td>本环节学时</td><td>300 分钟</td></tr>
<tr><td>典型工作过程描述</td><td colspan="4">用电安全规范—分析用电技术要求—勘察现场用电环境—确定选用方案—实施现场用电</td></tr>
<tr><td rowspan="6">调查项目</td><td>序号</td><td>调 查 内 容</td><td colspan="2">理由描述</td></tr>
<tr><td></td><td></td><td colspan="2"></td></tr>
<tr><td></td><td></td><td colspan="2"></td></tr>
<tr><td></td><td></td><td colspan="2"></td></tr>
<tr><td></td><td></td><td colspan="2"></td></tr>
<tr><td></td><td></td><td colspan="2"></td></tr>
<tr><td colspan="5">您对本次课程教学的改进意见：</td></tr>
<tr><td>调查信息</td><td>被调查人姓名</td><td></td><td>调查日期</td><td></td></tr>
</table>

1.4　电焊机临时用电学习情境分组单

学习场	金　工　实　训					
学习情境 1	电焊机临时用电				本环节学时	300 分钟
典型工作过程描述	用电安全规范—分析用电技术要求—勘察现场用电环境—确定选用方案—实施现场用电					
分组情况	组别	组长	组　员			
分组说明						
班级		教师签字		日期		

1.5　电焊机临时用电学习情境评价单

<table>
<tr><td>学习场</td><td colspan="7">金　工　实　训</td></tr>
<tr><td>学习情境 1</td><td colspan="4">电焊机临时用电</td><td colspan="2">本环节学时</td><td>20 分钟</td></tr>
<tr><td>典型工作过程描述</td><td colspan="7">用电安全规范—分析用电技术要求—勘察现场用电环境—确定选用方案—实施现场用电</td></tr>
<tr><td>评价项目</td><td colspan="3">评 价 子 项 目</td><td colspan="2">学生自评 20%</td><td>组内评价 30%</td><td>教师评价 50%</td></tr>
<tr><td>用电安全规范环节</td><td colspan="3">1. 资讯 10 分；2. 计划 10 分；3. 决策 10 分；4. 实施 40 分；5. 检查 10 分；6. 工作态度 20 分。</td><td colspan="2"></td><td></td><td></td></tr>
<tr><td>分析用电技术要求环节</td><td colspan="3">1. 资讯 10 分；2. 计划 10 分；3. 决策 10 分；4. 实施 40 分；5. 检查 10 分；6. 工作态度 20 分。</td><td colspan="2"></td><td></td><td></td></tr>
<tr><td>勘察现场用电环境环节</td><td colspan="3">1. 资讯 10 分；2. 计划 10 分；3. 决策 10 分；4. 实施 40 分；5. 检查 10 分；6. 工作态度 20 分。</td><td colspan="2"></td><td></td><td></td></tr>
<tr><td>确定选用方案环节</td><td colspan="3">1. 资讯 10 分；2. 计划 10 分；3. 决策 10 分；4. 实施 40 分；5. 检查 10 分；6. 工作态度 20 分。</td><td colspan="2"></td><td></td><td></td></tr>
<tr><td>实施现场用电环节</td><td colspan="3">1. 资讯 10 分；2. 计划 10 分；3. 决策 10 分；4. 实施 40 分；5. 检查 10 分；6. 工作态度 20 分。</td><td colspan="2"></td><td></td><td></td></tr>
<tr><td>最终结果</td><td colspan="3">1. 资讯 10 分；2. 计划 10 分；3. 决策 10 分；4. 实施 40 分；5. 检查 10 分；6. 工作态度 20 分。</td><td colspan="2"></td><td></td><td></td></tr>
<tr><td rowspan="3">评价</td><td>班级</td><td></td><td colspan="2">第（　　）组</td><td colspan="2">组长签字</td><td></td></tr>
<tr><td>教师签字</td><td></td><td colspan="2">日期</td><td colspan="3"></td></tr>
<tr><td colspan="7">评语：</td></tr>
</table>

1.6　教师实施计划单

学习场	金　工　实　训					
学习情境 1	电焊机临时用电			学时	300 分钟	
典型工作过程描述	用电安全规范—分析用电技术要求—勘察现场用电环境—确定选用方案—实施现场用电					
序号	工作与学习步骤	学时	使用工具	地点	方式	备注
班级		教师签字		日期		

Learning field

Metalworking practical training

Learning scenario 1
Temporary electricity of electric welding machine

1.0 Learning worksheet

<table>
<tr><td>Learning field</td><td colspan="6">Metalworking practical training</td></tr>
<tr><td>Learning scenario 1</td><td colspan="3">Temporary electricity of electric welding machine</td><td colspan="2">Total teaching time</td><td>300 minutes</td></tr>
<tr><td>Description of the typical working processes</td><td colspan="6">Electricity safety code – analysing electricity technical requirements – on-site survey on the environment of electricity utilization – confirming the scheme selection – implementing on-site electricity utilization</td></tr>
<tr><td>Learning objectives</td><td colspan="6">1.Master the temporary electricity safety code on the construction site.
2.Master the electricity technical requirements for electric welding machine on the construction site.
3.Master the selection of electric welding machines on the construction site.
4.Establish professional awareness of electricity utilization safety and cultivate the dedication to work.</td></tr>
<tr><td>Task description</td><td colspan="6">A construction site needs to weld the lightning protection earthing electrode (wire). Now, in accordance with the requirements, you have to connect the temporary power supply to the electric welding machine. Please complete the task safely and effectively according to the on-site situation and requirements, you can refer to Figure 1.0.1.</td></tr>
<tr><td>Teaching schedule</td><td>Information: 30 minutes</td><td>Planning: 30 minutes</td><td>Decision: 30 minutes</td><td>Implementation: 160 minutes</td><td>Check: 30 minutes</td><td>Assessment: 20 minutes</td></tr>
<tr><td>Requirements for students</td><td colspan="6">Preview the practical training homework sheets before taking the class and use the Internet to find the learning materials related to our practical training. Wear the training uniform throughout the training process. Implement the “4 No-harm Principles” (do not harm yourself, do not harm others, do not get harmed by others, protect others from being harmed) and other safety requirements.Strictly follow the rules and regulations of the training room.</td></tr>
<tr><td>References</td><td colspan="6">Refer to The General Specification for Arc Welding Machines (GB/T 8118—2010), Technical Code for Safety of Temporary Electrification on Construction Site” (JGJ 46—2005), Arc Welding Equipment – Part 1: Welding Power Sources (GB/T 15579.1—2013), and Safety Code of Electric Power Industry — Electric Part of Power Plants and Transformer Substations (GB 26860—2011).</td></tr>
</table>

Figure 1.0.1 On-site temporary electricity utilization

1.1 Teaching guidance

Learning field	Metalworking practical training	Learning scenario 1	Temporary electricity of electric welding machine	Reference frame	*The General Specification for Arc Welding Machines* (GB/T 8118—2010)	
General working processes / Typical working processes	Information	Planning	Decision	Implementation	Check	Assessment
Electricity safety code	Check the *Technical Code for safefy of Temporary Electrification on Construction Site* (JGJ 46–2005) to obtain relevant information about the electricity safety code on the construction site, so as to complete the tasks safely and effectively.	Through group discussions among students, prepare an electricity safety code scheme and complete 1.1.1.2 Planning sheet for electricity safety code.	Take integrated evaluation based on the feasibility, economy, operability, and implementation difficulty of the plans and select the best one.	Check relevant information and complete 1.1.1.4 Implementation sheet for electricity safety code.	Conduct self-check by students and check by the teacher according to the check items and check criteria included in 1.1.1.5 Check sheet for electricity safety code.	Conduct self-assessment, peer assessment, and teacher assessment according to the points assigned to the evaluation items and complete 1.5 Assessment sheet for the temporary electricity of electric welding machine learning scenario.

Continue

Learning field	Metalworking practical training	Learning scenario 1	Temporary electricity of electric welding machine	Reference frame	*The General Specification for Arc Welding Machines* (GB/T 8118—2010)	
General working processes / Typical working processes	Information	Planning	Decision	Implementation	Check	Assessment
Analysing electricity technical requirements	Check the *Arc Welding Equipment — Part 1: Welding Power Sources* (GB/T15579.1—2013) to obtain relevant information about the welding electricity technical requirements on the construction site, so as to complete the tasks safely and effectively.	Through group discussions among students, prepare an electricity safety code scheme and complete 1.1.2.2 Planning sheet for analysing electricity technical requirements.	Take integrated evaluation based on the feasibility, economy, operability, and implementation difficulty of the plan and select the best plan.	Check relevant information and complete 1.1.2.4 Implementation sheet for analysing electricity technical requirements.	Conduct self-check by students and check by the teacher according to the check items and check criteria included in 1.1.2.5 Check sheet for analysing electricity technical requirements.	Conduct self-assessment, peer assessment, and teacher assessment according to the points assigned to the evaluation items and complete 1.5 Assessment sheet for the temporary electricity of electric welding machine learning scenario.
On-site survey on the environment of electricity utilization	Check the *Technical Code for Safety of Temporary Electrification on Construction Site* (JGJ 46—2005) to obtain relevant information about the electricity safety code on the construction site, so as to complete the tasks safely and effectively.	Through group discussions among students, prepare an electricity safety code scheme and complete 1.1.3.2 Planning sheet for on-site survey on the environment of electricity utilization.	Take integrated evaluation based on the feasibility, economy, operability, and implementation difficulty of the plans and select the best one.	Check relevant information and complete 1.1.3.4 Implementation sheet for on-site survey on the environment of electricity utilization.	Conduct self-check by students and check by the teacher according to the check items and check criteria included in 1.1.3.5 Check sheet for on-site survey on the environment of electricity utilization.	Conduct self-assessment, peer assessment, and teacher assessment according to the points assigned to the evaluation items and complete 1.5 Assessment sheet for the temporary electricity of electric welding machine learning scenario.

Continue

Learning field	Metalworking practical training	Learning scenario 1	Temporary electricity of electric welding machine	Reference frame	*The General Specification for Arc Welding Machines* (GB/T 8118—2010)	
General working processes / Typical working processes	Information	Planning	Decision	Implementation	Check	Assessment
Confirming the scheme selection	Check the *Technical Code for Safety of Temporary Electrification on Construction Site* (JGJ 46—2005) to obtain relevant information about the electricity safety code on the construction site, so as to complete the tasks safely and effectively.	Through group discussions among students, prepare an electricity safety code scheme and complete 1.1.4.2 Planning sheet for confirming the scheme selection.	Take integrated evaluation based on the feasibility, economy, operability, and implementation difficulty of the plans and select the best one.	Check relevant information and complete 1.1.4.4 Implementation sheet for confirming the scheme selection.	Conduct self-check by students and check by the teacher according to the check items and check criteria included in 1.1.4.5 Check sheet for confirming the scheme selection.	Conduct self-assessment, peer assessment, and teacher assessment according to the points assigned to the evaluation items and complete 1.5 Assessment sheet for the temporary electricity for electric welding machine learning scenario.
Implementing on-site electricity utilization	Check the *Arc Welding Equipment—Part 1: Welding Power Sources* (GB/T 15579.1—2013) to obtain relevant information about the welding electricity technical requirements on the construction site, so as to complete the tasks safely and effectively.	Through group discussions among students, prepare an electricity safety code scheme and complete 1.1.5.2 Planning sheet for implementing on-site electricity utilization.	Take integrated evaluation based on the feasibility, economy, operability, and implementation difficulty of the plans and select the best one.	Check relevant information and complete 1.1.5.4 Implementation sheet for implementing on-site electricity utilization.	Conduct self-check by students and check by the teacher according to the check items and check criteria included in 1.1.5.5 Check sheet for implementing on-site electricity utilization.	Conduct self-assessment, peer assessment, and teacher assessment according to the points assigned to the evaluation items and complete 1.5 Assessment sheet for the temporary electricity of electric welding machine learning scenario.

1.1.1 Electricity safety code

1.1.1.1 Information sheet for electricity safety code

Learning field	Metalworking practical training		
Learning scenario 1	Temporary electricity of electric welding machine	Session teaching time	6 minutes
Description of the typical working processes	Electricity safety code – analysing electricity technical requirements – on-site survey on the environment of electricity utilization – confirming the scheme selection – implementing on-site electricity utilization		
Information collection methods	Teacher's explanation, search on the Internet, or communication among students		
Information description	Review the materials provided by the instructor to obtain information about the on-site electricity safety code, so as to complete the tasks safely and effectively, as indicated in Figure 1.1.1.  Figure 1.1.1 On-site survey		
Requirements for students	Preview the practical training homework sheets before taking the class and use the Internet to find the learning materials related to our practical training. Wear the training uniform throughout the training process. Implement the "4 No-harm Principles" (do not harm yourself, do not harm others, do not get harmed by others, protect others from being harmed) and other safety requirements.Strictly follow the rules and regulations of the training room.		
References	Refer to *Safety Code of Electric Power Industry — Electric Part of Power Plants And Transformer Substations* (GB 26860—2011), *Technical Code for Safety of Temporary Electrification on Construction Site* (JGJ 46—2005), and *Arc Welding Equipment — Part 1: Welding Power Sources*" (GB/T 15579.1—2013).		

1.1.1.2 Planning sheet for electricity safety code

<table>
<tr><td>Learning field</td><td colspan="5">Metalworking practical training</td></tr>
<tr><td>Learning scenario 1</td><td colspan="2">Temporary electricity of electric welding machine</td><td colspan="2">Session teaching time</td><td>6 minutes</td></tr>
<tr><td>Description of the typical working processes</td><td colspan="5">Electricity safety code – analysing electricity technical requirements – on-site survey on the environment of electricity utilization – confirming the scheme selection – implementing on-site electricity utilization</td></tr>
<tr><td>Planning methods</td><td colspan="5">Group discussion among classmates</td></tr>
<tr><td>Number</td><td>Working steps</td><td colspan="4">Notice</td></tr>
<tr><td>1</td><td>Conduct pre-class electricity safety education</td><td colspan="4">Check whether the class members are in good spirits, understand their mastery of the electricity safety code, cultivate their occupational safety awareness and teamwork spirit, and establish their professional values and dedication to work.</td></tr>
<tr><td>2</td><td>Watch the video about the on-site rescue procedures for electric shock</td><td colspan="4"></td></tr>
<tr><td>3</td><td>Learn about the electricity safety knowledge on the construction site, relevant national industry standards, and the essence of relevant documents</td><td colspan="4"></td></tr>
<tr><td>4</td><td>Discuss the learning benefits</td><td colspan="4"></td></tr>
<tr><td>5</td><td>Analyse the handling of emergency accidents such as electric shock or fire</td><td colspan="4"></td></tr>
<tr><td rowspan="3">Plan assessment</td><td>Class</td><td></td><td>Group ( )</td><td>Group leader's signature</td><td></td></tr>
<tr><td>Teacher's signature</td><td></td><td>Date</td><td colspan="2"></td></tr>
<tr><td colspan="5">Comments:</td></tr>
</table>

1.1.1.3 Decision sheet for electricity safety code

<table>
<tr><td>Learning field</td><td colspan="5">Metalworking practical training</td></tr>
<tr><td>Learning scenario 1</td><td colspan="3">Temporary electricity of electric welding machine</td><td>Session teaching time</td><td>6 minutes</td></tr>
<tr><td>Description of the typical working processes</td><td colspan="5">Electricity safety code – analysing electricity technical requirements – on-site survey on the environment of electricity utilization – confirming the scheme selection – implementing on-site electricity utilization</td></tr>
<tr><td colspan="6">Plan comparison</td></tr>
<tr><td>Number</td><td>Feasibility</td><td>Economy</td><td>Operability</td><td>Implementation difficulty</td><td>Integrated evaluation</td></tr>
<tr><td>Plan 1</td><td></td><td></td><td></td><td></td><td></td></tr>
<tr><td>Plan 2</td><td></td><td></td><td></td><td></td><td></td></tr>
<tr><td>Plan 3</td><td></td><td></td><td></td><td></td><td></td></tr>
<tr><td rowspan="3">Decision assessment</td><td>Class</td><td></td><td>Group ( )</td><td>Group leader's signature</td><td></td></tr>
<tr><td>Teacher's signature</td><td></td><td>Date</td><td colspan="2"></td></tr>
<tr><td colspan="5">Comments:</td></tr>
</table>

1.1.1.4 Implementation sheet for electricity safety code

<table>
<tr><td>Learning field</td><td colspan="6">Metalworking practical training</td></tr>
<tr><td>Learning scenario 1</td><td colspan="2">Temporary electricity of electric welding machine</td><td colspan="3">Session teaching time</td><td>30 minutes</td></tr>
<tr><td>Description of the typical working processes</td><td colspan="6">Electricity safety code – analysing electricity technical requirements – on-site survey on the environment of electricity utilization – confirming the scheme selection – implementing on-site electricity utilization</td></tr>
<tr><td>Number</td><td>Implementation steps</td><td colspan="3">Operations and methods</td><td colspan="2">Notice</td></tr>
<tr><td>1</td><td>Conduct a pre-class meeting</td><td colspan="3">Conduct a pre-class meeting on electricity safety education, check whether the class members are in good spirits, understand their mastery of the electricity safety code, cultivate their occupational safety awareness and teamwork spirit, and establish their professional values and dedication to work.</td><td colspan="2"></td></tr>
<tr><td>2</td><td>Watch the video</td><td colspan="3">Watch the video about the on-site rescue procedures for electric shock.</td><td colspan="2">Rescue skills for electric shock</td></tr>
<tr><td>3</td><td>Learn about the documents and standards</td><td colspan="3">Learn about the Safety Code of Electric Power Industry — Electric Part of Power Plants and Transformer Substations (GB 26860—2011) and Technical Code for Safety of Temporary Electrification on Construction Site (JGJ 46—2005). Search and study the diagram of the temporary electricity distribution box system, etc.</td><td colspan="2"></td></tr>
<tr><td>4</td><td>Discuss the learning benefits</td><td colspan="3"></td><td colspan="2"></td></tr>
<tr><td>5</td><td>Analyse the handling of emergency accidents</td><td colspan="3"></td><td colspan="2"></td></tr>
<tr><td colspan="7">Implementation instructions:</td></tr>
<tr><td rowspan="3">Implementation assessment</td><td>Class</td><td></td><td>Group ( )</td><td colspan="2">Group leader's signature</td><td></td></tr>
<tr><td>Teacher's signature</td><td></td><td>Date</td><td colspan="3"></td></tr>
<tr><td colspan="6">Comments:</td></tr>
</table>

1.1.1.5 Check sheet for electricity safety code

<table>
<tr><td>Learning field</td><td colspan="6">Metalworking practical training</td></tr>
<tr><td>Learning scenario 1</td><td colspan="4">Temporary electricity of electric welding machine</td><td>Session teaching time</td><td>6 minutes</td></tr>
<tr><td>Description of the typical working processes</td><td colspan="6">Electricity safety code – analysing electricity technical requirements – on-site survey on the environment of electricity utilization – confirming the scheme selection – implementing on-site electricity utilization</td></tr>
<tr><td>Number</td><td colspan="2">Check items</td><td colspan="2">Check criteria</td><td>Self-check by students</td><td>Check by the teacher</td></tr>
<tr><td>1</td><td colspan="2">Electricity safety code information</td><td colspan="2">Obtain the source of information about the on-site electricity safety code and identify whether it is the electrical safety code and standard of the electrical industry.</td><td></td><td></td></tr>
<tr><td>2</td><td colspan="2">Electricity safety code planning</td><td colspan="2">Formulate relevant plans for electricity safety code according to existing conditions and specific situations of the construction site.</td><td></td><td></td></tr>
<tr><td>3</td><td colspan="2">Electricity safety code decision</td><td colspan="2">Compare scores of various electricity safety code plans and select the best solution.</td><td></td><td></td></tr>
<tr><td>4</td><td colspan="2">Electricity safety code implementation</td><td colspan="2">Operate each step of the electricity safety code in accordance with the national industry standards and strictly implement the “4 No-harm Principles”.</td><td></td><td></td></tr>
<tr><td>5</td><td colspan="2">Electricity safety code check</td><td colspan="2">Check each step carefully, and check whether the theme idea of dedication to work is followed.</td><td></td><td></td></tr>
<tr><td rowspan="3">Check assessment</td><td>Class</td><td colspan="2"></td><td>Group ( )</td><td>Group leader’s signature</td><td></td></tr>
<tr><td>Teacher’s signature</td><td colspan="2"></td><td>Date</td><td colspan="2"></td></tr>
<tr><td colspan="6">Comments:</td></tr>
</table>

1.1.2 Analysing electricity technical requirements

1.1.2.1 Information sheet for analysing electricity technical requirements

Learning field	Metalworking practical training		
Learning scenario 1	Temporary electricity of electric welding machine	Session teaching time	6 minutes
Description of the typical working processes	Electricity safety code – analysing electricity technical requirements – on-site survey on the environment of electricity utilization – confirming the scheme selection – implementing on-site electricity utilization		
Information collection methods	Teacher's explanation, search on the Internet, or communication among students		
Information description	Review the materials provided by the instructor to obtain information about the on-site electricity safety code for the electric welding machine, so as to complete the tasks safely and effectively, as indicated in Figure 1.1.2.  Figure 1.1.2 Search for information		
Requirements for students	Preview the practical training homework sheets before taking the class and use the Internet to find the learning materials related to our practical training. Wear the training uniform throughout the training process. Implement the "4 No-harm Principles" (do not harm yourself, do not harm others, do not get harmed by others, protect others from being harmed) and other safety requirements. Strictly follow the rules and regulations of the training room.		
References	Refer to *Safety Code of Electric Power Industry–Electric Part of Power Plants and Transformer Substations* (GB 26860—2011) and *Technical Code for Safety of Temporary Electrification on Construction Site* (JGJ 46—2005).		

1.1.2.2 Planning sheet for analysing electricity technical requirements

Learning field	Metalworking practical training		
Learning scenario 1	Temporary electricity of electric welding machine	Session teaching time	6 minutes
Description of the typical working processes	Electricity safety code – analysing electricity technical requirements – on-site survey on the environment of electricity utilization – confirming the scheme selection – implementing on-site electricity utilization		
Planning methods	Group discussion among classmates		

Number	Working steps	Notice
1	Check the parameters of the electric welding machine provided on site	Check whether the electric welding machine uses single-phase alternating current or three-phase alternating current, and also check its current adjustment range.
2	Analyse whether the normal power of the electric welding machine fulfils the technical welding requirements	
3	Test the insulation performance of the electric welding machine	1. Conduct a self-check of the insulation megger, including short-circuit test and open-circuit test. 2. Maintain the rotational speed of the insulation megger at 120 r/min. 3. Pay attention to the constant speed of the insulation megger, then connect the "L" end. 4. Wait for the pointer of the insulation megger to stabilize before reading. 5. After the test, remove the "L" end , then stop shaking.
4	Analyse whether the working current of the distribution box air circuit breaker matches the working current of the electric welding machine	
5	Analyse whether the power cord is in line with the code and criteria	

Continue

<table>
<tr><td>Number</td><td>Working steps</td><td colspan="5">Notice</td></tr>
<tr><td>6</td><td>Analyse whether the installation of earthing lead of the electric welding machine is in line with the requirements</td><td colspan="5"></td></tr>
<tr><td rowspan="3">Plan assessment</td><td>Class</td><td></td><td>Group ( )</td><td>Group leader's signature</td><td colspan="2"></td></tr>
<tr><td>Teacher's signature</td><td></td><td>Date</td><td colspan="3"></td></tr>
<tr><td colspan="6">Comments:</td></tr>
</table>

1.1.2.3 Decision sheet for analysing electricity technical requirements

<table>
<tr><td>Learning field</td><td colspan="5">Metalworking practical training</td></tr>
<tr><td>Learning scenario 1</td><td colspan="3">Temporary electricity of electric welding machine</td><td>Session teaching time</td><td>6 minutes</td></tr>
<tr><td>Description of the typical working processes</td><td colspan="5">Electricity safety code – analysing electricity technical requirements – on-site survey on the environment of electricity utilization – confirming the scheme selection – implementing on-site electricity utilization</td></tr>
<tr><td colspan="6">Plan comparison</td></tr>
<tr><td>Number</td><td>Feasibility</td><td>Economy</td><td>Operability</td><td>Implementation difficulty</td><td>Integrated evaluation</td></tr>
<tr><td>Plan 1</td><td></td><td></td><td></td><td></td><td></td></tr>
<tr><td>Plan 2</td><td></td><td></td><td></td><td></td><td></td></tr>
<tr><td>Plan 3</td><td></td><td></td><td></td><td></td><td></td></tr>
<tr><td rowspan="3">Decision assessment</td><td>Class</td><td></td><td>Group ( )</td><td>Group leader's signature</td><td></td></tr>
<tr><td>Teacher's signature</td><td></td><td>Date</td><td colspan="2"></td></tr>
<tr><td colspan="5">Comments:</td></tr>
</table>

1.1.2.4 Implementation sheet for analysing electricity technical requirements

Learning field	Metalworking practical training		
Learning scenario 1	Temporary electricity of electric welding machine	Session teaching time	30 minutes
Description of the typical working processes	Electricity safety code – analysing electricity technical requirements – on-site survey on the environment of electricity utilization – confirming the scheme selection – implementing on-site electricity utilization		
Number	Implementation steps	Operations and methods	Notice
1	Check the parameters of the electric welding machine provided on site		
2	Analyse whether the normal power of the electric welding machine fulfils the technical welding requirements		
3	Test the insulation performance of the electric welding machine	1. Conduct a self-check of the insulation megger, including short-circuit test and open-circuit test. 2. Test the ground insulation resistance value of the primary side coil: connect the “E” end to the electric welding machine shell, and maintain the insulation megger’s rotational speed at 120 r/min; then connect the “L” end to the primary side coil and wait for the insulation megger’s pointer to stabilize before reading. After the test, remove the “L” end, then stop shaking. 3. Test the ground insulation resistance value of the secondary side coil and the value between primary and secondary side coils in the same way. Test whether the value is in line with the national standard of the electric power industry.	
4	Analyse whether the working current of the distribution box air circuit breaker matches the working current of the electric welding machine		
5	Analyse whether the power cord is in line with the code and criteria		

<table>
<tr><th>Number</th><th colspan="2">Implementation steps</th><th colspan="3">Operations and methods</th><th>Notice</th></tr>
<tr><td>6</td><td colspan="2">Analyse whether the installation of earthing lead of the electric welding machine is in line with the requirements</td><td colspan="3"></td><td></td></tr>
<tr><td colspan="7">Implementation instructions:</td></tr>
<tr><td rowspan="3">Implementation assessment</td><td>Class</td><td colspan="2"></td><td>Group ( )</td><td>Group leader's signature</td><td></td></tr>
<tr><td>Teacher's signature</td><td colspan="2"></td><td>Date</td><td colspan="2"></td></tr>
<tr><td colspan="6">Comments:</td></tr>
</table>

1.1.2.5 Check sheet for analysing electricity technical requirements

<table>
<tr><td>Learning field</td><td colspan="5">Metalworking practical training</td></tr>
<tr><td>Learning scenario 1</td><td colspan="2">Temporary electricity of electric welding machine</td><td colspan="2">Session teaching time</td><td>6 minutes</td></tr>
<tr><td>Description of the typical working processes</td><td colspan="5">Electricity safety code – analysing electricity technical requirements – on-site survey on the environment of electricity utilization – confirming the scheme selection – implementing on-site electricity utilization</td></tr>
<tr><td>Number</td><td>Check items</td><td colspan="2">Check criteria</td><td>Self-check by students</td><td>Check by the teacher</td></tr>
<tr><td>1</td><td>Information</td><td colspan="2">Obtain the source of information about the on-site electricity safety code of the electric welding machine, and identify whether it fulfils the national standard of the electric power industry.</td><td></td><td></td></tr>
<tr><td>2</td><td>Planning</td><td colspan="2">Formulate relevant plans that fulfils the electricity utilization of the electric welding machine on the construction site.</td><td></td><td></td></tr>
<tr><td>3</td><td>Decision</td><td colspan="2">Compare scores of various electricity utilization plans of the electric welding machine and select the best solution.</td><td></td><td></td></tr>
<tr><td>4</td><td>Implementation</td><td colspan="2">Operate each step of the electricity utilization of the electric welding machine in accordance with the standards, and strictly implement the “4 No-harm Principles”.</td><td></td><td></td></tr>
<tr><td>5</td><td>Check</td><td colspan="2">Check each step carefully, and check whether the theme idea of dedication to work is followed.</td><td></td><td></td></tr>
<tr><td rowspan="3">Check assessment</td><td>Class</td><td></td><td>Group ( )</td><td>Group leader’s signature</td><td></td></tr>
<tr><td>Teacher’s signature</td><td></td><td>Date</td><td colspan="2"></td></tr>
<tr><td colspan="5">Comments:</td></tr>
</table>

1.1.3 On-site survey on the environment of electricity utilization

1.1.3.1 Information sheet for the on-site survey on the environment of electricity utilization

<table>
<tr><td>Learning field</td><td colspan="3">Metalworking practical training</td></tr>
<tr><td>Learning scenario 1</td><td>Temporary electricity of electric welding machine</td><td>Session teaching time</td><td>6 minutes</td></tr>
<tr><td>Description of the typical working processes</td><td colspan="3">Electricity safety code – analysing electricity technical requirements – on-site survey on the environment of electricity utilization – confirming the scheme selection – implementing on-site electricity utilization</td></tr>
<tr><td>Information collection methods</td><td colspan="3">Teacher's explanation, search on the Internet, or communication among students</td></tr>
<tr><td>Information description</td><td colspan="3">Review the materials provided by the teacher to obtain information about the on-site electricity utilization environment of the electric welding machine, so as to complete the tasks safely and effectively, as indicated in Figure 1.1.3.

Figure 1.1.3 On-site survey on the environment of electricity utilization</td></tr>
<tr><td>Requirements for students</td><td colspan="3">Preview the practical training homework sheets before taking the class and use the Internet to find the learning materials related to our practical training. Wear the training uniform throughout the training process. Implement the "4 No-harm Principles" (do not harm yourself, do not harm others, do not get harmed by others, protect others from being harmed) and other safety requirements. Strictly follow the rules and regulations of the training room.</td></tr>
<tr><td>References</td><td colspan="3">Refer to Safety Code of Electric Power Industry — Electric Part of Power Plants and Transformer Substations (GB 26860 — 2011) and Technical Code for Safety of Temporary Electrification on Construction Site (JGJ 46 — 2005).</td></tr>
</table>

1.1.3.2 Planning sheet for the on-site survey on the environment of electricity utilization

<table>
<tr><td>Learning field</td><td colspan="6">Metalworking practical training</td></tr>
<tr><td>Learning scenario 1</td><td colspan="4">Temporary electricity of electric welding machine</td><td>Session teaching time</td><td>6 minutes</td></tr>
<tr><td>Description of the typical working processes</td><td colspan="6">Electricity safety code – analysing electricity technical requirements – on-site survey on the environment of electricity utilization – confirming the scheme selection – implementing on-site electricity utilization</td></tr>
<tr><td>Planning methods</td><td colspan="6">Group discussion among classmates</td></tr>
<tr><td>Number</td><td colspan="2">Working steps</td><td colspan="4">Notice</td></tr>
<tr><td>1</td><td colspan="2">Preparation before entering the construction site</td><td colspan="4">Check whether you are neatly dressed and wearing a helmet and other personal protective equipment correctly.</td></tr>
<tr><td>2</td><td colspan="2">Enter the construction site</td><td colspan="4"></td></tr>
<tr><td>3</td><td colspan="2">On-site survey work</td><td colspan="4">Check whether adequate fire equipment is set up at the welding work area; Check whether the distance between the welding work area and the storage and usage of the oxygen cylinder, calcium carbide barrel, and acetylene generator meets the prescribed safety distance; Check whether the distance between the electricity source and the welding work area meets the prescribed safety distance; Check whether the power switch is equipped with the leak-proof electricity device; Check whether the surface of the welding work area is in line with the requirements; Check whether other protective measures are in place at the site.</td></tr>
<tr><td>4</td><td colspan="2">Leave the construction site</td><td colspan="4"></td></tr>
<tr><td rowspan="3">Plan assessment</td><td>Class</td><td colspan="2"></td><td>Group ( )</td><td>Group leader's signature</td><td></td></tr>
<tr><td>Teacher's signature</td><td colspan="2"></td><td>Date</td><td colspan="2"></td></tr>
<tr><td colspan="6">Comments:</td></tr>
</table>

1.1.3.3 Decision sheet for the on-site survey on the environment of electricity utilization

<table>
<tr><td>Learning field</td><td colspan="5">Metalworking practical training</td></tr>
<tr><td>Learning scenario 1</td><td colspan="3">Temporary electricity of electric welding machine</td><td>Session teaching time</td><td>6 minutes</td></tr>
<tr><td>Description of the typical working processes</td><td colspan="5">Electricity safety code – analysing electricity technical requirements – on-site survey on the environment of electricity utilization – confirming the scheme selection – implementing on-site electricity utilization</td></tr>
<tr><td colspan="6">Plan comparison</td></tr>
<tr><td>Number</td><td>Feasibility</td><td>Economy</td><td>Operability</td><td>Implementation difficulty</td><td>Integrated evaluation</td></tr>
<tr><td>Plan 1</td><td></td><td></td><td></td><td></td><td></td></tr>
<tr><td>Plan 2</td><td></td><td></td><td></td><td></td><td></td></tr>
<tr><td>Plan 3</td><td></td><td></td><td></td><td></td><td></td></tr>
<tr><td rowspan="3">Decision assessment</td><td>Class</td><td></td><td>Group ( )</td><td>Group leader's signature</td><td></td></tr>
<tr><td>Teacher's signature</td><td></td><td>Date</td><td colspan="2"></td></tr>
<tr><td colspan="5">Comments:</td></tr>
</table>

1.1.3.4 Implementation sheet for on-site survey on the environment of electricity utilization

Learning field	Metalworking practical training		
Learning scenario 1	Temporary electricity of electric welding machine	Session teaching time	30 minutes
Description of the typical working processes	Electricity safety code – analysing electricity technical requirements – on-site survey on the environment of electricity utilization – confirming the scheme selection – implementing on-site electricity utilization		
Number	Implementation steps	Operations and methods	Notice
1	Preparation before entering the construction site	Check whether you are neatly dressed and wearing a helmet and other personal protective equipment correctly.	
2	Enter the construction site		
3	On-site survey work	Check whether adequate fire equipment is set up at the welding work area; Check whether the distance between the welding work area and the storage and usage of the oxygen cylinder, calcium carbide barrel, and acetylene generator meets the prescribed safety distance; Check whether the distance between the electricity source and the welding work area meets the prescribed safety distance; Check whether the power switch is equipped with the leak-proof electricity device; Check whether the surface of the welding work area is in line with the requirements; Check whether other protective measures are in place at the site.	
4	Leave the construction site		
Implementation instructions:			

Implementation assessment	Class		Group ( )	Group leader's signature	
	Teacher's signature		Date		
	Comments:				

1.1.3.5 Check sheet for on-site survey on the environment of electricity utilization

<table>
<tr><td colspan="2">Learning field</td><td colspan="5">Metalworking practical training</td></tr>
<tr><td colspan="2">Learning scenario 1</td><td colspan="3">Temporary electricity of electric welding machine</td><td>Session teaching time</td><td>6 minutes</td></tr>
<tr><td colspan="2">Description of the typical working processes</td><td colspan="5">Electricity safety code – analysing electricity technical requirements – on-site survey on the environment of electricity utilization – confirming the scheme selection – implementing on-site electricity utilization</td></tr>
<tr><td>Number</td><td colspan="2">Check items</td><td colspan="2">Check criteria</td><td>Self-check by students</td><td>Check by the teacher</td></tr>
<tr><td>1</td><td colspan="2">Information</td><td colspan="2">Obtain the source of information about the on-site survey on the electricity utilization environment of the electric welding machine.</td><td></td><td></td></tr>
<tr><td>2</td><td colspan="2">Planning</td><td colspan="2">Formulate relevant plans for on-site survey on the electricity utilization environment of the electric welding machine.</td><td></td><td></td></tr>
<tr><td>3</td><td colspan="2">Decision</td><td colspan="2">Compare scores of various on-site survey plans and select the best solution.</td><td></td><td></td></tr>
<tr><td>4</td><td colspan="2">Implementation</td><td colspan="2">Operate each step of the on-site survey on the environment of electricity utilization in accordance with the standards, and strictly implement the “4 No-harm Principles”.</td><td></td><td></td></tr>
<tr><td>5</td><td colspan="2">Check</td><td colspan="2">Check each step carefully, and check whether the theme idea of dedication to work is followed.</td><td></td><td></td></tr>
<tr><td colspan="2" rowspan="3">Check assessment</td><td>Class</td><td></td><td>Group ( )</td><td>Group leader’s signature</td><td></td></tr>
<tr><td>Teacher’s signature</td><td></td><td>Date</td><td colspan="2"></td></tr>
<tr><td colspan="5">Comments:</td></tr>
</table>

1.1.4 Confirming the scheme selection

1.1.4.1 Information sheet for confirming the scheme selection

Learning field	Metalworking practical training		
Learning scenario 1	Temporary electricity of electric welding machine	Session teaching time	6 minutes
Description of the typical working processes	Electricity safety code – analysing electricity technical requirements – on-site survey on the environment of electricity utilization – confirming the scheme selection – implementing on-site electricity utilization		
Information collection methods	Teacher's explanation, search on the Internet, or communication among students		
Information description	Review the materials provided by the teacher to obtain information about the on-site electric welding machine scheme, so as to complete the tasks safely and effectively, as indicated in Figure 1.1.4.  Figure 1.1.4 Confirm the scheme selection		
Requirements for students	Preview the practical training homework sheets before taking the class and use the Internet to find the learning materials related to our practical training.Wear the training uniform throughout the training process.Implement the "4 No-harm Principles" (do not harm yourself, do not harm others, do not get harmed by others, protect others from being harmed) and other safety requirements.Strictly follow the rules and regulations of the training room.		
References	Refer to *Safety Code of Electric Power Industry — Electric Part of Power Plants and Transformer Substations* (GB 26860—2011) and "*Technical Code for Safety of Temporary Electrification on Construction Site* (JGJ 46—2005).		

1.1.4.2 Planning sheet for confirming the scheme selection

<table>
<tr><td>Learning field</td><td colspan="6">Metalworking practical training</td></tr>
<tr><td>Learning scenario 1</td><td colspan="3">Temporary electricity of electric welding machine</td><td colspan="2">Session teaching time</td><td>6 minutes</td></tr>
<tr><td>Description of the typical working processes</td><td colspan="6">Electricity safety code – analysing electricity technical requirements – on-site survey on the environment of electricity utilization – confirming the scheme selection – implementing on-site electricity utilization</td></tr>
<tr><td>Planning methods</td><td colspan="6">Group discussion among classmates</td></tr>
<tr><td>Number</td><td colspan="2">Working steps</td><td colspan="4">Notice</td></tr>
<tr><td>1</td><td colspan="2">Inquire about the national standards of the electric power industry</td><td colspan="4"></td></tr>
<tr><td>2</td><td colspan="2">Compose the scheme</td><td colspan="4"></td></tr>
<tr><td>3</td><td colspan="2">Scheme submission</td><td colspan="4"></td></tr>
<tr><td>4</td><td colspan="2">Group discussion</td><td colspan="4"></td></tr>
<tr><td>5</td><td colspan="2">Confirm the scheme</td><td colspan="4"></td></tr>
<tr><td rowspan="3">Plan assessment</td><td>Class</td><td></td><td>Group ( )</td><td>Group leader's signature</td><td colspan="2"></td></tr>
<tr><td>Teacher's signature</td><td></td><td>Date</td><td colspan="3"></td></tr>
<tr><td colspan="6">Comments:</td></tr>
</table>

1.1.4.3 Decision sheet for confirming the scheme selection

<table>
<tr><td>Learning field</td><td colspan="5">Metalworking practical training</td></tr>
<tr><td>Learning scenario 1</td><td colspan="3">Temporary electricity of electric welding machine</td><td>Session teaching time</td><td>6 minutes</td></tr>
<tr><td>Description of the typical working processes</td><td colspan="5">Electricity safety code – analysing electricity technical requirements – on-site survey on the environment of electricity utilization – confirming the scheme selection – implementing on-site electricity utilization</td></tr>
<tr><td colspan="6">Plan comparison</td></tr>
<tr><td>Number</td><td>Feasibility</td><td>Economy</td><td>Operability</td><td>Implementation difficulty</td><td>Integrated evaluation</td></tr>
<tr><td>Plan 1</td><td></td><td></td><td></td><td></td><td></td></tr>
<tr><td>Plan 2</td><td></td><td></td><td></td><td></td><td></td></tr>
<tr><td>Plan 3</td><td></td><td></td><td></td><td></td><td></td></tr>
<tr><td rowspan="3">Decision assessment</td><td>Class</td><td></td><td>Group ( )</td><td>Group leader's signature</td><td></td></tr>
<tr><td>Teacher's signature</td><td></td><td>Date</td><td colspan="2"></td></tr>
<tr><td colspan="5">Comments:</td></tr>
</table>

1.1.4.4 Implementation sheet for confirming the scheme selection

<table>
<tr><td>Learning field</td><td colspan="5">Metalworking practical training</td></tr>
<tr><td>Learning scenario 1</td><td colspan="2">Temporary electricity of electric welding machine</td><td colspan="2">Session teaching time</td><td>30 minutes</td></tr>
<tr><td>Description of the typical working processes</td><td colspan="5">Electricity safety code – analysing electricity technical requirements – on-site survey on the environment of electricity utilization – confirming the scheme selection – implementing on-site electricity utilization</td></tr>
<tr><td>Number</td><td>Implementation steps</td><td colspan="2">Operations and methods</td><td colspan="2">Notice</td></tr>
<tr><td>1</td><td>Inquire about the related standards</td><td colspan="2"></td><td colspan="2"></td></tr>
<tr><td>2</td><td>Compose the scheme</td><td colspan="2"></td><td colspan="2"></td></tr>
<tr><td>3</td><td>Scheme submission</td><td colspan="2"></td><td colspan="2"></td></tr>
<tr><td>4</td><td>Group discussion</td><td colspan="2"></td><td colspan="2"></td></tr>
<tr><td>5</td><td>Confirm the scheme</td><td colspan="2"></td><td colspan="2"></td></tr>
<tr><td colspan="6">Implementation instructions:</td></tr>
</table>

<table>
<tr><td rowspan="3">Implementation assessment</td><td>Class</td><td></td><td>Group ( )</td><td>Group leader's signature</td><td></td></tr>
<tr><td>Teacher's signature</td><td></td><td>Date</td><td colspan="2"></td></tr>
<tr><td colspan="5">Comments:</td></tr>
</table>

1.1.4.5 Check sheet for confirming the scheme selection

<table>
<tr><td>Learning field</td><td colspan="4">Metalworking practical training</td></tr>
<tr><td>Learning scenario 1</td><td colspan="2">Temporary electricity of electric welding machine</td><td>Session teaching time</td><td>6 minutes</td></tr>
<tr><td>Description of the typical working processes</td><td colspan="4">Electricity safety code – analysing electricity technical requirements – on-site survey on the environment of electricity utilization – confirming the scheme selection – implementing on-site electricity utilization</td></tr>
</table>

Number	Check items	Check criteria	Self-check by students	Check by the teacher
1	Scheme selection information	Obtain the source of information about the on-site electric welding machine selection.		
2	Scheme selection planning	Formulate relevant plans for electric welding machine selection on the construction site.		
3	Scheme selection decision	Compare scores of various electric welding machine selection schemes on the construction site and select the best solution.		
4	Scheme selection implementation	Operate each step of on-site electric welding machine selection in accordance with the standards, and strictly implement the “4 No-harm Principles”.		
5	Scheme selection check	Check each step carefully, and check whether the theme idea of dedication to work is followed.		

<table>
<tr><td rowspan="3">Check assessment</td><td>Class</td><td></td><td>Group ( )</td><td>Group leader’s signature</td><td></td></tr>
<tr><td>Teacher’s signature</td><td></td><td>Date</td><td colspan="2"></td></tr>
<tr><td colspan="5">Comments:</td></tr>
</table>

1.1.5 Implementing on-site electricity utilization

1.1.5.1 Information sheet for implementing on-site electricity utilization

<table>
<tr><td>Learning field</td><td colspan="3">Metalworking practical training</td></tr>
<tr><td>Learning scenario 1</td><td>Temporary electricity of electric welding machine</td><td>Session teaching time</td><td>6 minutes</td></tr>
<tr><td>Description of the typical working processes</td><td colspan="3">Electricity safety code – analysing electricity technical requirements – on-site survey on the environment of electricity utilization – confirming the scheme selection – implementing on-site electricity utilization</td></tr>
<tr><td>Information collection methods</td><td colspan="3">Teacher's explanation, search on the Internet, or communication among students</td></tr>
<tr><td>Information description</td><td colspan="3">Review the materials provided by the instructor to obtain information about the on-site electricity utilization for electric welding machine, so as to complete the tasks safely and effectively, as indicated in Figure 1.1.5.

Figure 1.1.5 On-site electricity utilization</td></tr>
<tr><td>Requirements for students</td><td colspan="3">Preview the practical training homework sheets before taking the class and use the Internet to find the learning materials related to our practical training. Wear the training uniform throughout the training process. Implement the "4 No-harm Principles" (do not harm yourself, do not harm others, do not get harmed by others, protect others from being harmed) and other safety requirements. Strictly follow the rules and regulations of the training room.</td></tr>
<tr><td>References</td><td colspan="3">Refer to Safety Code of Electric Power Industry — Electric Part of Power Plants and Transformer Substations (GB 26860—2011) and Technical Code for Safety of Temporary Electrification on Construction Site (JGJ 46—2005).</td></tr>
</table>

1.1.5.2 Planning sheet for implementing on-site electricity utilization

<table>
<tr><td>Learning field</td><td colspan="5">Metalworking practical training</td></tr>
<tr><td>Learning scenario 1</td><td colspan="2">Temporary electricity of electric welding machine</td><td colspan="2">Session teaching time</td><td>6 minutes</td></tr>
<tr><td>Description of the typical working processes</td><td colspan="5">Electricity safety code – analysing electricity technical requirements – on-site survey on the environment of electricity utilization – confirming the scheme selection – implementing on-site electricity utilization</td></tr>
<tr><td>Planning methods</td><td colspan="5">Group discussion among classmates</td></tr>
<tr><td>Number</td><td>Working steps</td><td colspan="3">Operations</td><td>Notice</td></tr>
<tr><td>1</td><td>Preparation before the operation, implement on-site safety measures</td><td colspan="3">Check the spirits of the staff; Carry out work permit procedures; The person in charge should conduct a handover to the staff on on-site safety and specify the task, the work area, the parts with electricity, and situations of safety measures.</td><td></td></tr>
<tr><td>2</td><td>Electricity cut</td><td colspan="3"></td><td></td></tr>
<tr><td>3</td><td>Electricity check</td><td colspan="3"></td><td></td></tr>
<tr><td>4</td><td>Device installation, wiring</td><td colspan="3"></td><td></td></tr>
<tr><td>5</td><td>Electricity transmission</td><td colspan="3"></td><td></td></tr>
<tr><td>6</td><td>Welding</td><td colspan="3"></td><td></td></tr>
<tr><td rowspan="3">Plan assessment</td><td>Class</td><td></td><td>Group ( )</td><td>Group leader's signature</td><td></td></tr>
<tr><td>Teacher's signature</td><td></td><td>Date</td><td colspan="2"></td></tr>
<tr><td colspan="5">Comments:</td></tr>
</table>

1.1.5.3 Decision sheet for implementing on-site electricity utilization

<table>
<tr><td>Learning field</td><td colspan="5">Metalworking practical training</td></tr>
<tr><td>Learning scenario 1</td><td colspan="3">Temporary electricity of electric welding machine</td><td>Session teaching time</td><td>6 minutes</td></tr>
<tr><td>Description of the typical working processes</td><td colspan="5">Electricity safety code – analysing electricity technical requirements – on-site survey on the environment of electricity utilization – confirming the scheme selection – implementing on-site electricity utilization</td></tr>
<tr><td colspan="6">Plan comparison</td></tr>
<tr><td>Number</td><td>Feasibility</td><td>Economy</td><td>Operability</td><td>Implementation difficulty</td><td>Integrated evaluation</td></tr>
<tr><td>Plan 1</td><td></td><td></td><td></td><td></td><td></td></tr>
<tr><td>Plan 2</td><td></td><td></td><td></td><td></td><td></td></tr>
<tr><td>Plan 3</td><td></td><td></td><td></td><td></td><td></td></tr>
<tr><td rowspan="3">Decision assessment</td><td>Class</td><td></td><td>Group ( )</td><td>Group leader’s signature</td><td></td></tr>
<tr><td>Teacher’s signature</td><td></td><td>Date</td><td colspan="2"></td></tr>
<tr><td colspan="5">Comments:</td></tr>
</table>

1.1.5.4 Implementation sheet for implementing on-site electricity utilization

<table>
<tr><td>Learning field</td><td colspan="4">Metalworking practical training</td></tr>
<tr><td>Learning scenario 1</td><td colspan="2">Temporary electricity of electric welding machine</td><td>Session teaching time</td><td>40 minutes</td></tr>
<tr><td>Description of the typical working processes</td><td colspan="4">Electricity safety code – analysing electricity technical requirements – on-site survey on the environment of electricity utilization – confirming the scheme selection – implementing on-site electricity utilization</td></tr>
<tr><td>Number</td><td>Implementation steps</td><td>Operations and methods</td><td colspan="2">Notice</td></tr>
<tr><td>1</td><td>Preparation before the operation, implement on-site safety measures</td><td>Check the spirits of the staff; Carry out work permit procedures; The person in charge should conduct a handover to the staff on on-site safety and specify the task, the work area, the parts with electricity, and situations of safety measures.</td><td colspan="2">Prevention of electric shock accidents</td></tr>
<tr><td>2</td><td>Electricity cut</td><td>Implement the rule of "one person operating while one person supervising". The operator should wear a helmet and insulated gloves and pull down the power switch.</td><td colspan="2">Prevention of electric shock accidents</td></tr>
<tr><td>3</td><td>Electricity check</td><td></td><td colspan="2"></td></tr>
<tr><td>4</td><td>Device installation, wiring</td><td></td><td colspan="2"></td></tr>
<tr><td>5</td><td>Electricity transmission</td><td></td><td colspan="2"></td></tr>
<tr><td>6</td><td>Welding</td><td></td><td colspan="2"></td></tr>
<tr><td colspan="5">Implementation instructions:</td></tr>
</table>

<table>
<tr><td rowspan="3">Implementation assessment</td><td>Class</td><td></td><td>Group ( )</td><td>Group leader's signature</td><td></td></tr>
<tr><td>Teacher's signature</td><td></td><td>Date</td><td></td><td></td></tr>
<tr><td colspan="5">Comments:</td></tr>
</table>

1.1.5.5 Check sheet for implementing on-site electricity utilization

<table>
<tr><td colspan="2">Learning field</td><td colspan="4">Metalworking practical training</td></tr>
<tr><td colspan="2">Learning scenario 1</td><td colspan="2">Temporary electricity of electric welding machine</td><td>Session teaching time</td><td>6 minutes</td></tr>
<tr><td colspan="2">Description of the typical working processes</td><td colspan="4">Electricity safety code – analysing electricity technical requirements – on-site survey on the environment of electricity utilization – confirming the scheme selection – implementing on-site electricity utilization</td></tr>
<tr><td>Number</td><td>Check items</td><td>Check criteria</td><td>Self-check by students</td><td colspan="2">Check by the teacher</td></tr>
<tr><td>1</td><td>Information</td><td>Obtain the source of information about the on-site electricity utilization of the electric welding machine.</td><td></td><td colspan="2"></td></tr>
<tr><td>2</td><td>Planning</td><td>Formulate relevant plans for electricity utilization of the electric welding machine on the construction site.</td><td></td><td colspan="2"></td></tr>
<tr><td>3</td><td>Decision</td><td>Compare scores of various plans for the electricity utilization of the electric welding machine on the construction site and select the best solution.</td><td></td><td colspan="2"></td></tr>
<tr><td>4</td><td>Implementation</td><td>Operate each step of the on-site electricity utilization of the electric welding machine in accordance with the standards and strictly implement the “4 No-harm Principles”.</td><td></td><td colspan="2"></td></tr>
<tr><td>5</td><td>Check</td><td>Check each step carefully, and check whether the theme idea of dedication to work is followed.</td><td></td><td colspan="2"></td></tr>
<tr><td rowspan="3">Check assessment</td><td>Class</td><td></td><td>Group ( )</td><td>Group leader’s signature</td><td></td></tr>
<tr><td>Teacher’s signature</td><td></td><td>Date</td><td></td><td></td></tr>
<tr><td colspan="5">Comments:</td></tr>
</table>

1.2 List of materials and tools for the temporary electricity of electric welding machine learning scenario

Learning field	Metalworking practical training				
Learning scenario 1	Temporary electricity of electric welding machine		Session teaching time	300 minutes	
Description of the typical working processes	Electricity safety code – analysing electricity technical requirements – on-site survey on the environment of electricity utilization – confirming the scheme selection – implementing on-site electricity utilization				
Number	Name	Function	Quantity	Model	User
Class		Group ( )	Group leader's signature		
Teacher's signature		Date			

1.3　Teaching feedback sheet for the temporary electricity of electric welding machine learning scenario

<table>
<tr><td>Learning field</td><td colspan="4">Metalworking practical training</td></tr>
<tr><td>Learning scenario 1</td><td colspan="2">Temporary electricity of electric welding machine</td><td>Session teaching time</td><td>300 minutes</td></tr>
<tr><td>Description of the typical working processes</td><td colspan="4">Electricity safety code – analysing electricity technical requirements – on-site survey on the environment of electricity utilization – confirming the scheme selection – implementing on-site electricity utilization</td></tr>
<tr><td rowspan="6">Survey items</td><td>Number</td><td>Survey contents</td><td colspan="2">Description of the reasons</td></tr>
<tr><td></td><td></td><td colspan="2"></td></tr>
<tr><td></td><td></td><td colspan="2"></td></tr>
<tr><td></td><td></td><td colspan="2"></td></tr>
<tr><td></td><td></td><td colspan="2"></td></tr>
<tr><td></td><td></td><td colspan="2"></td></tr>
<tr><td colspan="5">Your feedback on how to improve the teaching of this course:</td></tr>
<tr><td>Survey information</td><td>Respondent's name</td><td></td><td>Survey date</td><td></td></tr>
</table>

1.4 Grouping sheet for the temporary electricity of electric welding machine learning scenario

Learning field	Metalworking practical training				
Learning scenario 1	Temporary electricity of electric welding machine		Session teaching time	300 minutes	
Description of the typical working processes	Electricity safety code – analysing electricity technical requirements – on-site survey on the environment of electricity utilization – confirming the scheme selection – implementing on-site electricity utilization				
Grouping	Group	Group leader	Group members		
Grouping instructions					
Class		Teacher's signature		Date	

1.5 Assessment sheet for the temporary electricity of electric welding machine learning scenario

Learning field	Metalworking practical training			
Learning scenario 1	Temporary electricity of electric welding machine		Session teaching time	20 minutes
Description of the typical working processes	Electricity safety code – analysing electricity technical requirements – on-site survey on the environment of electricity utilization – confirming the scheme selection – implementing on-site electricity utilization			
Assessment items	Assessment subitems	Self-assessment by students 20%	Intra-group assessment 30%	Assessment by the teacher 50%
Electricity safety code	1. Information: 10 points; 2. Planning: 10 points; 3. Decision: 10 points; 4. Implementation: 40 points; 5. Check: 10 points; 6. Work attitude: 20 points.			
Analysing electricity technical requirements	1. Information: 10 points; 2. Planning: 10 points; 3. Decision: 10 points; 4. Implementation: 40 points; 5. Check: 10 points; 6. Work attitude: 20 points.			
On-site survey on the environment of electricity utilization	1. Information: 10 points; 2. Planning: 10 points; 3. Decision: 10 points; 4. Implementation: 40 points; 5. Check: 10 points; 6. Work attitude: 20 points.			
Confirming the scheme selection	1. Information: 10 points; 2. Planning: 10 points; 3. Decision: 10 points; 4. Implementation: 40 points; 5. Check: 10 points; 6. Work attitude: 20 points.			
Implementing on-site electricity utilization	1. Information: 10 points; 2. Planning: 10 points; 3. Decision: 10 points; 4. Implementation: 40 points; 5. Check: 10 points; 6. Work attitude: 20 points.			

Continue

<table>
<tr><th>Assessment items</th><th colspan="2">Assessment subitems</th><th>Self-assessment by students 20%</th><th>Intra-group assessment 30%</th><th>Assessment by the teacher 50%</th></tr>
<tr><td>Final results</td><td colspan="2">1. Information: 10 points;
2. Planning: 10 points;
3. Decision: 10 points;
4. Implementation: 40 points;
5. Check: 10 points;
6. Work attitude: 20 points.</td><td></td><td></td><td></td></tr>
<tr><td rowspan="3">Assessment</td><td>Class</td><td></td><td>Group ( )</td><td>Group leader's signature</td><td></td></tr>
<tr><td>Teacher's signature</td><td></td><td>Date</td><td colspan="2"></td></tr>
<tr><td colspan="5">Comments:</td></tr>
</table>

1.6　Teacher's implementation planning sheet

<table>
<tr><td>Learning field</td><td colspan="7">Metalworking practical training</td></tr>
<tr><td>Learning scenario 1</td><td colspan="3">Temporary electricity of electric welding machine</td><td colspan="2">Session teaching time</td><td colspan="2">300 minutes</td></tr>
<tr><td>Description of the typical working processes</td><td colspan="7">Electricity safety code – analysing electricity technical requirements – on-site survey on the environment of electricity utilization – confirming the scheme selection – implementing on-site electricity utilization</td></tr>
<tr><td>Number</td><td colspan="2">Working and learning steps</td><td>Teaching time</td><td>Tools</td><td>Locations</td><td>Methods</td><td>Notes</td></tr>
<tr><td></td><td colspan="2"></td><td></td><td></td><td></td><td></td><td></td></tr>
<tr><td></td><td colspan="2"></td><td></td><td></td><td></td><td></td><td></td></tr>
<tr><td></td><td colspan="2"></td><td></td><td></td><td></td><td></td><td></td></tr>
<tr><td></td><td colspan="2"></td><td></td><td></td><td></td><td></td><td></td></tr>
<tr><td></td><td colspan="2"></td><td></td><td></td><td></td><td></td><td></td></tr>
<tr><td></td><td colspan="2"></td><td></td><td></td><td></td><td></td><td></td></tr>
<tr><td></td><td colspan="2"></td><td></td><td></td><td></td><td></td><td></td></tr>
<tr><td></td><td colspan="2"></td><td></td><td></td><td></td><td></td><td></td></tr>
<tr><td></td><td colspan="2"></td><td></td><td></td><td></td><td></td><td></td></tr>
<tr><td></td><td colspan="2"></td><td></td><td></td><td></td><td></td><td></td></tr>
<tr><td>Class</td><td></td><td>Teacher's signature</td><td colspan="2"></td><td>Date</td><td colspan="2"></td></tr>
</table>

学习情境 2

台钻床临时用电

2.0 学习性工作任务单

<table>
<tr><td>学习场</td><td colspan="6">金　工　实　训</td></tr>
<tr><td>学习情境 2</td><td colspan="3">台钻床临时用电</td><td colspan="2">总学时</td><td>300 分钟</td></tr>
<tr><td>典型工作过程描述</td><td colspan="6">用电安全规范—分析用电技术要求—勘察现场用电环境—确定选用方案—实施现场用电</td></tr>
<tr><td>学习目标</td><td colspan="6">1. 掌握施工现场临时用电规范。
2. 掌握施工现场台钻床用电技术要求。
3. 掌握施工现场台钻床的选用。
4. 树立安全用电职业意识，培养爱岗敬业精神。</td></tr>
<tr><td>任务描述</td><td colspan="6">某施工现场需要给 220kV 线夹板加工钻孔，现要求按照规范接上临时电源供给一台钻床钻孔使用。请你根据现场具体情况，安全有效地完成任务，参考图 2.0.1。

图 2.0.1　现场用电</td></tr>
<tr><td>学时安排</td><td>资讯
30 分钟</td><td>计划
30 分钟</td><td>决策
30 分钟</td><td>实施
160 分钟</td><td>检查
30 分钟</td><td>评价
20 分钟</td></tr>
<tr><td>对学生的要求</td><td colspan="6">课前充分预习实训作业单，充分利用网络查找有关实训的学习资料，实训过程中穿着实训服装，贯彻落实自己不伤害自己、自己不伤害他人、自己不被他人伤害、保护他人不被伤害的“四不伤害”安全要求，注意严格遵守实训室的各项规章制度。</td></tr>
<tr><td>参考资料</td><td colspan="6">参考《电力安全工作规程　发电厂和变电站电气部分》（GB 26860—2011）和《施工现场临时用电安全技术规范》（JGJ 46—2005）。</td></tr>
</table>

2.1 教学引导文

学习场	金工实训	学习情境 2	台钻床临时用电	参照系	《施工现场临时用电安全技术规范》(JGJ 46—2005)	
普适性工作过程 / 典型工作过程	资讯	计划	决策	实施	检查	评价
用电安全规范	查看《施工现场临时用电安全技术规范》(JGJ 46—2005),获取施工现场用电规范的有关信息,便于安全有效地完成任务。	通过同学之间小组讨论,拟订用电安全规范计划,完成 2.1.1.2 用电安全规范环节计划单。	计划的可行性、经济性、可操作性和实施难度几方面综合分析,选取最佳方案。	查找有关资料,完成 2.1.1.4 用电安全规范环节实施单。	按照 2.1.1.5 用电安全规范环节检查单的检查项目和标准,进行学生自查和教师检查。	根据评价项目配置分值,进行自评、互评和教师评价,完成 2.5 台钻床临时用电学习情境评价单。
分析用电技术要求	查看《施工现场临时用电安全技术规范》(JGJ 46—2005),获取施工现场点焊接用电技术要求的有关信息,便于安全有效地完成任务。	通过同学之间小组讨论,拟订用电安全规范计划,完成 2.1.2.2 分析用电技术要求环节计划单。	计划的可行性、经济性、可操作性和实施难度几方面综合分析,选取最佳方案。	查找有关资料,完成 2.1.2.4 分析用电技术要求环节实施单。	按照 2.1.2.5 分析用电技术要求环节检查单的检查项目和标准,进行学生自查和教师检查。	根据评价项目配置分值,进行自评、互评和教师评价,完成 2.5 台钻床临时用电学习情境评价单。
勘察现场用电环境	查看《施工现场临时用电安全技术规范》(JGJ 46—2005),获取施工现场用电规范的有关信息,便于安全有效地完成任务。	通过同学之间小组讨论,拟订用电安全规范计划,完成 2.1.3.2 勘察现场用电环境环节计划单。	计划的可行性、经济性、可操作性和实施难度几方面综合分析,选取最佳方案。	查找有关资料,完成 2.1.3.4 勘察现场用电环境环节实施单。	按照 2.1.3.5 勘察现场用电环境环节检查单的检查项目和标准,进行学生自查和教师检查。	根据评价项目配置分值,进行自评、互评和教师评价,完成 2.5 台钻床临时用电学习情境评价单。

续表

学习场	金工实训	学习情境2	台钻床临时用电	参照系	《施工现场临时用电安全技术规范》(JGJ 46—2005)	
普适性工作过程 典型工作过程	资讯	计划	决策	实施	检查	评价
确定选用方案	查看《施工现场临时用电安全技术规范》(JGJ 46—2005)，获取施工现场用电规范的有关信息，便于安全有效地完成任务。	通过同学之间小组讨论，拟订用电安全规范计划，完成2.1.4.2确定选用方案环节计划单。	计划的可行性、经济性、可操作性和实施难度几方面综合分析，选取最佳方案。	查找有关资料，完成2.1.4.4确定选用方案环节实施单。	按照2.1.4.5确定选用方案环节检查单的检查项目和标准，进行学生自查和教师检查。	根据评价项目配置分值，进行自评、互评和教师评价，完成2.5台钻床临时用电学习情境评价单。
实施现场用电	查看《施工现场临时用电安全技术规范》(JGJ 46—2005)，获取施工现场点焊接用电技术要求的有关信息，便于安全有效地完成任务。	通过同学之间小组讨论，拟订用电安全规范计划，完成2.1.5.2实施现场用电环节计划单。	计划的可行性、经济性、可操作性和实施难度几方面综合分析，选取最佳方案。	查找有关资料，完成2.1.5.4实施现场用电环节实施单。	按照2.1.5.5实施现场用电环节检查单的检查项目和标准，进行学生自查和教师检查。	根据评价项目配置分值，进行自评、互评和教师评价，完成2.5台钻床临时用电学习情境评价单。

2.1.1　用电安全规范

2.1.1.1　用电安全规范环节资讯单

<table>
<tr><td>学习场</td><td colspan="3">金　工　实　训</td></tr>
<tr><td>学习情境 2</td><td>台钻床临时用电</td><td>本环节学时</td><td>6 分钟</td></tr>
<tr><td>典型工作过程描述</td><td colspan="3">用电安全规范—分析用电技术要求—勘察现场用电环境—确定选用方案—实施现场用电</td></tr>
<tr><td>搜集资讯的方式</td><td colspan="3">教师的讲解、互联网查询或者同学间交流</td></tr>
<tr><td>资讯描述</td><td colspan="3">查看教师提供的资料，获取台钻床用电规范的有关信息，便于安全有效地完成任务，如图 2.1.1 所示。

图 2.1.1　现场查勘</td></tr>
<tr><td>对学生的要求</td><td colspan="3">课前充分预习实训作业单，充分利用网络查找有关实训的学习资料，实训过程中穿着实训服装，贯彻落实自己不伤害自己、自己不伤害他人、自己不被他人伤害、保护他人不被伤害的“四不伤害”安全要求，注意严格遵守实训室的各项规章制度。</td></tr>
<tr><td>参考资料</td><td colspan="3">参考《电力安全工作规程　发电厂和变电站电气部分》(GB 26860—2011) 和《施工现场临时用电安全技术规范》(JGJ 46—2005)。</td></tr>
</table>

学习情境 2-4

2.1.1.2 用电安全规范环节计划单

<table>
<tr><td>学习场</td><td colspan="5">金 工 实 训</td></tr>
<tr><td>学习情境 2</td><td colspan="3">台钻床临时用电</td><td>本环节学时</td><td>6 分钟</td></tr>
<tr><td>典型工作过程描述</td><td colspan="5">用电安全规范—分析用电技术要求—勘察现场用电环境—确定选用方案—实施现场用电</td></tr>
<tr><td>计划制订的方式</td><td colspan="5">同学之间小组讨论</td></tr>
<tr><td>序号</td><td colspan="2">工 作 步 骤</td><td colspan="3">注 意 事 项</td></tr>
<tr><td>1</td><td colspan="2">开班前会进行用电安全教育</td><td colspan="3"></td></tr>
<tr><td>2</td><td colspan="2">观看现场触电救护视频</td><td colspan="3"></td></tr>
<tr><td>3</td><td colspan="2">学习施工现场用电安全知识和有关文件精神</td><td colspan="3"></td></tr>
<tr><td>4</td><td colspan="2">讨论学习的收获</td><td colspan="3"></td></tr>
<tr><td>5</td><td colspan="2">应急事故处理分析</td><td colspan="3"></td></tr>
<tr><td rowspan="3">计划评价</td><td>班级</td><td></td><td>第（ ）组</td><td>组长签字</td><td></td></tr>
<tr><td>教师签字</td><td></td><td>日期</td><td colspan="2"></td></tr>
<tr><td colspan="5">评语：</td></tr>
</table>

学习情境 2-5

2.1.1.3 用电安全规范环节决策单

<table>
<tr><td>学习场</td><td colspan="5">金 工 实 训</td></tr>
<tr><td>学习情境 2</td><td colspan="3">台钻床临时用电</td><td>本环节学时</td><td>6 分钟</td></tr>
<tr><td>典型工作过程描述</td><td colspan="5">用电安全规范—分析用电技术要求—勘察现场用电环境—确定选用方案—实施现场用电</td></tr>
<tr><td colspan="6">计 划 对 比</td></tr>
<tr><td>序号</td><td>计划的可行性</td><td>计划的经济性</td><td>计划的可操作性</td><td>计划的实施难度</td><td>综合评价</td></tr>
<tr><td>方案 1</td><td></td><td></td><td></td><td></td><td></td></tr>
<tr><td>方案 2</td><td></td><td></td><td></td><td></td><td></td></tr>
<tr><td>方案 3</td><td></td><td></td><td></td><td></td><td></td></tr>
<tr><td rowspan="3">决策评价</td><td>班级</td><td></td><td>第（　　）组</td><td>组长签字</td><td></td></tr>
<tr><td>教师签字</td><td></td><td>日期</td><td colspan="2"></td></tr>
<tr><td colspan="5">评语：</td></tr>
</table>

2.1.1.4　用电安全规范环节实施单

<table>
<tr><td colspan="2">学习场</td><td colspan="4">金　工　实　训</td></tr>
<tr><td colspan="2">学习情境 2</td><td colspan="2">台钻床临时用电</td><td>本环节学时</td><td>30 分钟</td></tr>
<tr><td colspan="2">典型工作过程描述</td><td colspan="4">用电安全规范—分析用电技术要求—勘察现场用电环境—确定选用方案—实施现场用电</td></tr>
<tr><td>序号</td><td>实施步骤</td><td colspan="2">工作内容及方法</td><td colspan="2">注 意 事 项</td></tr>
<tr><td>1</td><td>开班前会</td><td colspan="2">开班前会进行用电安全教育。</td><td colspan="2"></td></tr>
<tr><td>2</td><td>观看视频</td><td colspan="2">观看现场触电救护视频。</td><td colspan="2">触电急救技能</td></tr>
<tr><td>3</td><td>学习文件标准</td><td colspan="2">学习《电力安全工作规程　发电厂和变电站电气部分》（GB 26860—2011）和《施工现场临时用电安全技术规范》（JGJ 46—2005）。查找并学习临时用电配电箱系统图等。</td><td colspan="2"></td></tr>
<tr><td>4</td><td>谈学习收获</td><td colspan="2"></td><td colspan="2"></td></tr>
<tr><td>5</td><td>应急事故处理分析</td><td colspan="2"></td><td colspan="2"></td></tr>
<tr><td colspan="6">实施说明：</td></tr>
<tr><td rowspan="3">实施评价</td><td>班级</td><td></td><td>第（　　）组</td><td>组长签字</td><td></td></tr>
<tr><td>教师签字</td><td></td><td>日期</td><td colspan="2"></td></tr>
<tr><td colspan="5">评语：</td></tr>
</table>

2.1.1.5 用电安全规范环节检查单

学习场	金 工 实 训			
学习情境2	台钻床临时用电		本环节学时	6分钟
典型工作过程描述	用电安全规范—分析用电技术要求—勘察现场用电环境—确定选用方案—实施现场用电			
序号	检查项目	检 查 标 准	学生自查	教师检查
1	资讯环节	获得施工现场台钻床用电规范的有关信息的来源。		
2	计划环节	制定施工现场台钻床用电规范步骤的各个方案。		
3	决策环节	对各种施工现场台钻床用电规范的方案衡量对比评分，获取最佳方案。		
4	实施环节	施工现场台钻床用电规范过程中各个环节按照标准执行，贯彻落实“四不伤害”。		
5	检查环节	各个环节逐一检查，是否有爱岗敬业的主题思想。		
检查评价	班级		第（ ）组	组长签字
	教师签字		日期	
	评语：			

2.1.2　分析用电技术要求

2.1.2.1　分析用电技术要求环节资讯单

<table>
<tr><td>学习场</td><td colspan="3">金　工　实　训</td></tr>
<tr><td>学习情境 2</td><td>台钻床临时用电</td><td>本环节学时</td><td>6 分钟</td></tr>
<tr><td>典型工作过程描述</td><td colspan="3">用电安全规范—分析用电技术要求—勘察现场用电环境—确定选用方案—实施现场用电</td></tr>
<tr><td>搜集资讯的方式</td><td colspan="3">教师的讲解、互联网查询或者同学间交流</td></tr>
<tr><td>资讯描述</td><td colspan="3">查看教师提供的资料，获取台钻床安全用电技术要求的有关信息，便于安全有效地完成任务，如图 2.1.2 所示。

图 2.1.2　查看资料</td></tr>
<tr><td>对学生的要求</td><td colspan="3">课前充分预习实训作业单，充分利用网络查找有关实训的学习资料，实训过程中穿着实训服装，贯彻落实自己不伤害自己、自己不伤害他人、自己不被他人伤害、保护他人不被伤害的“四不伤害”安全要求，注意严格遵守实训室的各项规章制度。</td></tr>
<tr><td>参考资料</td><td colspan="3">参考《电力安全工作规程　发电厂和变电站电气部分》（GB 26860—2011）和《施工现场临时用电安全技术规范》（JGJ 46—2005）。</td></tr>
</table>

2.1.2.2　分析用电技术要求环节计划单

<table>
<tr><td colspan="2">学习场</td><td colspan="4">金　工　实　训</td></tr>
<tr><td colspan="2">学习情境 2</td><td colspan="2">台钻床临时用电</td><td>本环节学时</td><td>6 分钟</td></tr>
<tr><td colspan="2">典型工作过程描述</td><td colspan="4">用电安全规范—分析用电技术要求—勘察现场用电环境—确定选用方案—实施现场用电</td></tr>
<tr><td colspan="2">计划制订的方式</td><td colspan="4">同学之间小组讨论</td></tr>
<tr><td>序号</td><td colspan="2">工 作 步 骤</td><td colspan="3">注 意 事 项</td></tr>
<tr><td>1</td><td colspan="2">查找现场的台钻床参数</td><td colspan="3">单相交流电还是三相交流电，电机功率大小，电动机通过五级变速带轮，主轴可否变转速，工作台能否在圆立柱—上下移动。</td></tr>
<tr><td>2</td><td colspan="2">分析台钻床正常功率是否满足现场钻孔的技术要求</td><td colspan="3"></td></tr>
<tr><td>3</td><td colspan="2">检测台钻床的绝缘性能</td><td colspan="3"></td></tr>
<tr><td>4</td><td colspan="2">分析配电箱空气开关工作电流是否与台钻床工作电流匹配</td><td colspan="3"></td></tr>
<tr><td>5</td><td colspan="2">分析台钻床电源线是否符合规范标准</td><td colspan="3"></td></tr>
<tr><td>6</td><td colspan="2">分析台钻床接地线的安装是否符合要求</td><td colspan="3"></td></tr>
<tr><td colspan="2" rowspan="3">计划评价</td><td>班级</td><td></td><td>第（　　）组</td><td>组长签字</td></tr>
<tr><td>教师签字</td><td></td><td>日期</td><td></td></tr>
<tr><td colspan="4">评语：</td></tr>
</table>

2.1.2.3 分析用电技术要求环节决策单

<table>
<tr><td>学习场</td><td colspan="6">金 工 实 训</td></tr>
<tr><td>学习情境 2</td><td colspan="4">台钻床临时用电</td><td>本环节学时</td><td>6 分钟</td></tr>
<tr><td>典型工作过程描述</td><td colspan="6">用电安全规范—分析用电技术要求—勘察现场用电环境—确定选用方案—实施现场用电</td></tr>
<tr><td colspan="7">计 划 对 比</td></tr>
<tr><td>序号</td><td>计划的可行性</td><td>计划的经济性</td><td>计划的可操作性</td><td>计划的实施难度</td><td colspan="2">综合评价</td></tr>
<tr><td>方案 1</td><td></td><td></td><td></td><td></td><td colspan="2"></td></tr>
<tr><td>方案 2</td><td></td><td></td><td></td><td></td><td colspan="2"></td></tr>
<tr><td>方案 3</td><td></td><td></td><td></td><td></td><td colspan="2"></td></tr>
<tr><td rowspan="3">决策评价</td><td>班级</td><td></td><td>第（ ）组</td><td>组长签字</td><td colspan="2"></td></tr>
<tr><td>教师签字</td><td></td><td>日期</td><td colspan="3"></td></tr>
<tr><td colspan="6">评语：</td></tr>
</table>

2.1.2.4 分析用电技术要求环节实施单

<table>
<tr><td>学习场</td><td colspan="5">金 工 实 训</td></tr>
<tr><td>学习情境 2</td><td colspan="3">台钻床临时用电</td><td>本环节学时</td><td>30 分钟</td></tr>
<tr><td>典型工作过程描述</td><td colspan="5">用电安全规范—分析用电技术要求—勘察现场用电环境—确定选用方案—实施现场用电</td></tr>
<tr><td>序号</td><td>实 施 步 骤</td><td colspan="3">工作内容及方法</td><td>注意事项</td></tr>
<tr><td>1</td><td>查找现场的台钻床参数</td><td colspan="3"></td><td></td></tr>
<tr><td>2</td><td>分析台钻床正常功率是否满足现场钻孔的技术要求</td><td colspan="3"></td><td></td></tr>
<tr><td>3</td><td>检测台钻床的绝缘性能</td><td colspan="3"></td><td></td></tr>
<tr><td>4</td><td>分析配电箱空气开关工作电流是否与台钻床工作电流匹配</td><td colspan="3"></td><td></td></tr>
<tr><td>5</td><td>分析台钻床电源线是否符合规范标准</td><td colspan="3"></td><td></td></tr>
<tr><td>6</td><td>分析台钻床接地线的安装是否符合要求</td><td colspan="3"></td><td></td></tr>
<tr><td colspan="6">实施说明：</td></tr>
<tr><td rowspan="3">实施评价</td><td>班级</td><td></td><td>第（ ）组</td><td>组长签字</td><td></td></tr>
<tr><td>教师签字</td><td></td><td>日期</td><td colspan="2"></td></tr>
<tr><td colspan="5">评语：</td></tr>
</table>

2.1.2.5　分析用电技术要求环节检查单

<table>
<tr><td colspan="2">学习场</td><td colspan="4">金　工　实　训</td></tr>
<tr><td colspan="2">学习情境2</td><td colspan="2">台钻床临时用电</td><td>本环节学时</td><td>6分钟</td></tr>
<tr><td colspan="2">典型工作过程描述</td><td colspan="4">用电安全规范—分析用电技术要求—勘察现场用电环境—确定选用方案—实施现场用电</td></tr>
<tr><td>序号</td><td>检查项目</td><td>检 查 标 准</td><td>学生自查</td><td colspan="2">教师检查</td></tr>
<tr><td>1</td><td>资讯环节</td><td>获得施工现场台钻床安全用电的有关信息的来源是否符合国家电力行业标准。</td><td></td><td colspan="2"></td></tr>
<tr><td>2</td><td>计划环节</td><td>制定施工现场台钻床安全用电步骤的各个方案。</td><td></td><td colspan="2"></td></tr>
<tr><td>3</td><td>决策环节</td><td>对各种施工现场台钻床安全用电的方案衡量对比评分，获取最佳方案。</td><td></td><td colspan="2"></td></tr>
<tr><td>4</td><td>实施环节</td><td>施工现场台钻床安全用电过程中各个环节按照标准执行，贯彻落实“四不伤害”。</td><td></td><td colspan="2"></td></tr>
<tr><td>5</td><td>检查环节</td><td>各个环节逐一检查，是否有爱岗敬业的主题思想。</td><td></td><td colspan="2"></td></tr>
</table>

<table>
<tr><td rowspan="3">检查评价</td><td>班级</td><td></td><td>第（　　）组</td><td>组长签字</td><td></td></tr>
<tr><td>教师签字</td><td></td><td>日期</td><td colspan="2"></td></tr>
<tr><td colspan="5">评语：</td></tr>
</table>

学习情境 2-13

2.1.3　勘察现场用电环境

2.1.3.1　勘察现场用电环境环节资讯单

学习场	金　工　实　训		
学习情境 2	台钻床临时用电	本环节学时	6 分钟
典型工作过程描述	用电安全规范—分析用电技术要求—勘察现场用电环境—确定选用方案—实施现场用电		
搜集资讯的方式	教师的讲解、互联网查询或者同学间交流		
资讯描述	查看教师提供的资料，获取施工现场台钻床安全用电的有关信息，便于安全有效地完成任务，如图 2.1.3 所示。  图 2.1.3　现场勘察环境		
对学生的要求	课前充分预习实训作业单，充分利用网络查找有关实训的学习资料，实训过程中穿着实训服装，贯彻落实自己不伤害自己、自己不伤害他人、自己不被他人伤害、保护他人不被伤害的“四不伤害”安全要求，注意严格遵守实训室的各项规章制度。		
参考资料	参考《电力安全工作规程　发电厂和变电站电气部分》(GB 26860—2011) 和《施工现场临时用电安全技术规范》(JGJ 46—2005)。		

2.1.3.2　勘察现场用电环境环节计划单

<table>
<tr><td colspan="2">学习场</td><td colspan="5">金　工　实　训</td></tr>
<tr><td colspan="2">学习情境 2</td><td colspan="3">台钻床临时用电</td><td>本环节学时</td><td>6 分钟</td></tr>
<tr><td colspan="2">典型工作过程描述</td><td colspan="5">用电安全规范—分析用电技术要求—勘察现场用电环境—确定选用方案—实施现场用电</td></tr>
<tr><td colspan="2">计划制订的方式</td><td colspan="5">同学之间小组讨论</td></tr>
<tr><td>序号</td><td>工 作 步 骤</td><td colspan="5">注 意 事 项</td></tr>
<tr><td>1</td><td>进入施工现场前准备工作</td><td colspan="5"></td></tr>
<tr><td>2</td><td>进入施工现场</td><td colspan="5"></td></tr>
<tr><td>3</td><td>现场勘察工作</td><td colspan="5">查看焊接作业点是否配置有足够数量的消防设备；查看焊接作业点与氧气瓶、电石桶和乙炔发生器在存放、使用过程中是否符合规定的安全距离；查看现有电源距离焊接作业点是否符合要求；查看电源开关是否配有漏电装置；查看焊接工作地面是否符合要求；查看现场其他防护措施是否到位。</td></tr>
<tr><td>4</td><td>离开施工现场</td><td colspan="5"></td></tr>
<tr><td colspan="2" rowspan="3">计划评价</td><td>班级</td><td></td><td>第（　　）组</td><td>组长签字</td><td></td></tr>
<tr><td>教师签字</td><td></td><td>日期</td><td colspan="2"></td></tr>
<tr><td colspan="5">评语：</td></tr>
</table>

学习情境 2-15

2.1.3.3　勘察现场用电环境环节决策单

<table>
<tr><td>学习场</td><td colspan="5">金　工　实　训</td></tr>
<tr><td>学习情境 2</td><td colspan="3">台钻床临时用电</td><td>本环节学时</td><td>6 分钟</td></tr>
<tr><td>典型工作过程描述</td><td colspan="5">用电安全规范—分析用电技术要求—勘察现场用电环境—确定选用方案—实施现场用电</td></tr>
<tr><td colspan="6">计　划　对　比</td></tr>
<tr><td>序号</td><td>计划的可行性</td><td>计划的经济性</td><td>计划的可操作性</td><td>计划的实施难度</td><td>综合评价</td></tr>
<tr><td>方案 1</td><td></td><td></td><td></td><td></td><td></td></tr>
<tr><td>方案 2</td><td></td><td></td><td></td><td></td><td></td></tr>
<tr><td>方案 3</td><td></td><td></td><td></td><td></td><td></td></tr>
</table>

<table>
<tr><td rowspan="3">决策评价</td><td>班级</td><td></td><td>第（　　）组</td><td>组长签字</td><td></td></tr>
<tr><td>教师签字</td><td></td><td>日期</td><td colspan="2"></td></tr>
<tr><td colspan="5">评语：</td></tr>
</table>

2.1.3.4　勘察现场用电环境环节实施单

<table>
<tr><td colspan="2">学习场</td><td colspan="4">金　工　实　训</td></tr>
<tr><td colspan="2">学习情境 2</td><td colspan="2">台钻床临时用电</td><td>本环节学时</td><td>30 分钟</td></tr>
<tr><td colspan="2">典型工作过程描述</td><td colspan="4">用电安全规范—分析用电技术要求—勘察现场用电环境—确定选用方案—实施现场用电</td></tr>
<tr><td>序号</td><td>实施步骤</td><td colspan="2">工作内容及方法</td><td colspan="2">注意事项</td></tr>
<tr><td>1</td><td>进入施工现场前准备工作</td><td colspan="2">检查着装是否整齐、是否戴好安全帽和其他个人安全防护用品等。</td><td colspan="2"></td></tr>
<tr><td>2</td><td>进入施工现场</td><td colspan="2"></td><td colspan="2"></td></tr>
<tr><td>3</td><td>现场勘察工作</td><td colspan="2">查看焊接作业点是否配置有足够数量的消防设备；查看焊接作业点与氧气瓶、电石桶和乙炔发生器在存放、使用过程中是否符合规定的安全距离；查看现有电源距离焊接作业点是否符合要求；查看电源开关是否配有漏电装置；查看焊接工作地面是否符合要求；查看现场其他防护措施是否到位。</td><td colspan="2"></td></tr>
<tr><td>4</td><td>离开施工现场</td><td colspan="2"></td><td colspan="2"></td></tr>
<tr><td colspan="6">实施说明：</td></tr>
<tr><td colspan="2" rowspan="3">实施评价</td><td>班级</td><td></td><td>第（　　）组</td><td>组长签字　</td></tr>
<tr><td>教师签字</td><td></td><td>日期</td><td></td></tr>
<tr><td colspan="4">评语：</td></tr>
</table>

学习情境 2-17

2.1.3.5　勘察现场用电环境环节检查单

<table>
<tr><td colspan="2">学习场</td><td colspan="6">金　工　实　训</td></tr>
<tr><td colspan="2">学习情境 2</td><td colspan="4">台钻床临时用电</td><td>本环节学时</td><td>6 分钟</td></tr>
<tr><td colspan="2">典型工作过程描述</td><td colspan="6">用电安全规范—分析用电技术要求—勘察现场用电环境—确定选用方案—实施现场用电</td></tr>
<tr><td>序号</td><td>检查项目</td><td colspan="4">检 查 标 准</td><td>学生自查</td><td>教师检查</td></tr>
<tr><td>1</td><td>资讯环节</td><td colspan="4">获得勘察施工现场台钻床用电环境的有关信息的来源是否符合国家电力行业的标准。</td><td></td><td></td></tr>
<tr><td>2</td><td>计划环节</td><td colspan="4">制定勘察施工现场台钻床用电环境步骤的各个方案。</td><td></td><td></td></tr>
<tr><td>3</td><td>决策环节</td><td colspan="4">对各种勘察施工现场台钻床用电环境的方案衡量对比评分，获取最佳方案。</td><td></td><td></td></tr>
<tr><td>4</td><td>实施环节</td><td colspan="4">勘察施工现场台钻床用电环境过程中各个环节按照标准执行，贯彻落实“四不伤害”。</td><td></td><td></td></tr>
<tr><td>5</td><td>检查环节</td><td colspan="4">各个环节逐一检查，是否有爱岗敬业的主题思想。</td><td></td><td></td></tr>
<tr><td colspan="2" rowspan="3">检查评价</td><td>班级</td><td></td><td>第（　　）组</td><td>组长签字</td><td colspan="2"></td></tr>
<tr><td>教师签字</td><td></td><td>日期</td><td colspan="3"></td></tr>
<tr><td colspan="6">评语：</td></tr>
</table>

2.1.4　确定选用方案

2.1.4.1　确定选用方案环节资讯单

学习场	金　工　实　训		
学习情境 2	台钻床临时用电	本环节学时	6 分钟
典型工作过程描述	用电安全规范—分析用电技术要求—勘察现场用电环境—确定选用方案—实施现场用电		
搜集资讯的方式	教师的讲解、互联网查询或者同学间交流		
资讯描述	查看教师提供的资料，获取施工现场选用台钻床的有关信息，便于安全有效地完成任务，如图 2.1.4 所示。  图 2.1.4　确定方案		
对学生的要求	课前充分预习实训作业单，充分利用网络查找有关实训的学习资料，实训过程中穿着实训服装，贯彻落实自己不伤害自己、自己不伤害他人、自己不被他人伤害、保护他人不被伤害的“四不伤害”安全要求，注意严格遵守实训室的各项规章制度。		
参考资料	参考《电力安全工作规程　发电厂和变电站电气部分》（GB 26860—2011）和《施工现场临时用电安全技术规范》（JGJ 46—2005）。		

2.1.4.2　确定选用方案环节计划单

<table>
<tr><td colspan="2">学习场</td><td colspan="5">金　工　实　训</td></tr>
<tr><td colspan="2">学习情境 2</td><td colspan="3">台钻床临时用电</td><td>本环节学时</td><td>6 分钟</td></tr>
<tr><td colspan="2">典型工作过程描述</td><td colspan="5">用电安全规范—分析用电技术要求—勘察现场用电环境—确定选用方案—实施现场用电</td></tr>
<tr><td colspan="2">计划制订的方式</td><td colspan="5">同学之间小组讨论</td></tr>
<tr><td>序号</td><td>工 作 步 骤</td><td colspan="5">注 意 事 项</td></tr>
<tr><td>1</td><td>查询有关标准</td><td colspan="5"></td></tr>
<tr><td>2</td><td>编写方案</td><td colspan="5"></td></tr>
<tr><td>3</td><td>提交方案</td><td colspan="5"></td></tr>
<tr><td>4</td><td>小组讨论</td><td colspan="5"></td></tr>
<tr><td>5</td><td>确定方案</td><td colspan="5"></td></tr>
<tr><td colspan="2" rowspan="3">计划评价</td><td>班级</td><td></td><td>第（　　）组</td><td>组长签字</td><td></td></tr>
<tr><td>教师签字</td><td></td><td>日期</td><td colspan="2"></td></tr>
<tr><td colspan="5">评语：</td></tr>
</table>

学习情境 2-20

2.1.4.3　确定选用方案环节决策单

<table>
<tr><td>学习场</td><td colspan="6">金　工　实　训</td></tr>
<tr><td>学习情境 2</td><td colspan="3">台钻床临时用电</td><td colspan="2">本环节学时</td><td>6 分钟</td></tr>
<tr><td>典型工作过程描述</td><td colspan="6">用电安全规范—分析用电技术要求—勘察现场用电环境—确定选用方案—实施现场用电</td></tr>
<tr><td colspan="7">计　划　对　比</td></tr>
<tr><td>序号</td><td>计划的可行性</td><td>计划的经济性</td><td colspan="2">计划的可操作性</td><td>计划的实施难度</td><td>综合评价</td></tr>
<tr><td>方案 1</td><td></td><td></td><td colspan="2"></td><td></td><td></td></tr>
<tr><td>方案 2</td><td></td><td></td><td colspan="2"></td><td></td><td></td></tr>
<tr><td>方案 3</td><td></td><td></td><td colspan="2"></td><td></td><td></td></tr>
<tr><td rowspan="3">决策评价</td><td>班级</td><td></td><td>第（　　）组</td><td>组长签字</td><td colspan="2"></td></tr>
<tr><td>教师签字</td><td></td><td>日期</td><td colspan="3"></td></tr>
<tr><td colspan="6">评语：</td></tr>
</table>

2.1.4.4 确定选用方案环节实施单

<table>
<tr><td colspan="2">学习场</td><td colspan="3">金　工　实　训</td></tr>
<tr><td colspan="2">学习情境 2</td><td>台钻床临时用电</td><td>本环节学时</td><td>30 分钟</td></tr>
<tr><td colspan="2">典型工作过程描述</td><td colspan="3">用电安全规范—分析用电技术要求—勘察现场用电环境—确定选用方案—实施现场用电</td></tr>
<tr><td>序号</td><td>实施步骤</td><td>工作内容及方法</td><td colspan="2">注意事项</td></tr>
<tr><td>1</td><td>查询有关标准</td><td></td><td colspan="2"></td></tr>
<tr><td>2</td><td>编写方案</td><td></td><td colspan="2"></td></tr>
<tr><td>3</td><td>提交方案</td><td></td><td colspan="2"></td></tr>
<tr><td>4</td><td>小组讨论</td><td></td><td colspan="2"></td></tr>
<tr><td>5</td><td>确定方案</td><td></td><td colspan="2"></td></tr>
<tr><td colspan="5">实施说明：</td></tr>
</table>

<table>
<tr><td rowspan="3">实施评价</td><td>班级</td><td></td><td>第（　　）组</td><td>组长签字</td><td></td></tr>
<tr><td>教师签字</td><td></td><td>日期</td><td colspan="2"></td></tr>
<tr><td colspan="5">评语：</td></tr>
</table>

学习情境 2-22

不经历风雨，怎么见彩虹。苦练千百次，操作如有神。

2.1.4.5 确定选用方案环节检查单

<table>
<tr><td>学习场</td><td colspan="5">金 工 实 训</td></tr>
<tr><td>学习情境 2</td><td colspan="3">台钻床临时用电</td><td>本环节学时</td><td>6 分钟</td></tr>
<tr><td>典型工作过程描述</td><td colspan="5">用电安全规范—分析用电技术要求—勘察现场用电环境—确定选用方案—实施现场用电</td></tr>
<tr><td>序号</td><td>检查项目</td><td colspan="2">检 查 标 准</td><td>学生自查</td><td>教师检查</td></tr>
<tr><td>1</td><td>资讯环节</td><td colspan="2">获得施工现场确定选用台钻床方案的有关信息的来源是否符合国家电力行业的标准。</td><td></td><td></td></tr>
<tr><td>2</td><td>计划环节</td><td colspan="2">制定施工现场确定选用台钻床方案步骤的各个方案。</td><td></td><td></td></tr>
<tr><td>3</td><td>决策环节</td><td colspan="2">对各种施工现场确定选用台钻床方案的方案衡量对比评分，获取最佳方案。</td><td></td><td></td></tr>
<tr><td>4</td><td>实施环节</td><td colspan="2">施工现场确定选用台钻床方案过程中各个环节按照标准执行，贯彻落实“四不伤害”。</td><td></td><td></td></tr>
<tr><td>5</td><td>检查环节</td><td colspan="2">各个环节逐一检查，是否有爱岗敬业的主题思想。</td><td></td><td></td></tr>
<tr><td rowspan="3">检查评价</td><td>班级</td><td></td><td>第（ ）组</td><td>组长签字</td><td></td></tr>
<tr><td>教师签字</td><td></td><td>日期</td><td colspan="2"></td></tr>
<tr><td colspan="5">评语：</td></tr>
</table>

2.1.5　实施现场用电

2.1.5.1　实施现场用电环节资讯单

学习场	金　工　实　训		
学习情境 2	台钻床临时用电	本环节学时	6 分钟
典型工作过程描述	用电安全规范—分析用电技术要求—勘察现场用电环境—确定选用方案—实施现场用电		
搜集资讯的方式	教师的讲解、互联网查询或者同学间交流		
资讯描述	查看教师提供的资料，获取台钻床施工现场用电的有关信息，便于安全有效地完成任务，如图 2.1.5 所示。  图 2.1.5　现场用电		
对学生的要求	课前充分预习实训作业单，充分利用网络查找有关实训的学习资料，实训过程中穿着实训服装，贯彻落实自己不伤害自己、自己不伤害他人、自己不被他人伤害、保护他人不被伤害的“四不伤害”安全要求，注意严格遵守实训室的各项规章制度。		
参考资料	参考《电力安全工作规程　发电厂和变电站电气部分》（GB 26860—2011）和《施工现场临时用电安全技术规范》（JGJ 46—2005）。		

2.1.5.2　实施现场用电环节计划单

<table>
<tr><td>学习场</td><td colspan="5">金　工　实　训</td></tr>
<tr><td>学习情境 2</td><td colspan="2">台钻床临时用电</td><td colspan="2">本环节学时</td><td>6 分钟</td></tr>
<tr><td>典型工作过程描述</td><td colspan="5">用电安全规范—分析用电技术要求—勘察现场用电环境—确定选用方案—实施现场用电</td></tr>
<tr><td>计划制订的方式</td><td colspan="5">同学之间小组讨论</td></tr>
<tr><td>序号</td><td>工 作 步 骤</td><td colspan="3">工 作 内 容</td><td>注意事项</td></tr>
<tr><td>1</td><td>作业前准备，落实现场安全措施</td><td colspan="3">检查工作人员的精神状态；办理工作许可手续；工作负责人进行现场安全交底，向工作人员指明工作任务、工作范围、带电部位及安全措施情况。</td><td></td></tr>
<tr><td>2</td><td>停电</td><td colspan="3"></td><td></td></tr>
<tr><td>3</td><td>验电</td><td colspan="3"></td><td></td></tr>
<tr><td>4</td><td>安装设备、接线</td><td colspan="3"></td><td></td></tr>
<tr><td>5</td><td>送电</td><td colspan="3"></td><td></td></tr>
<tr><td>6</td><td>钻床的操作</td><td colspan="3"></td><td></td></tr>
<tr><td rowspan="3">计划评价</td><td>班级</td><td></td><td>第（　　）组</td><td>组长签字</td><td></td></tr>
<tr><td>教师签字</td><td></td><td>日期</td><td colspan="2"></td></tr>
<tr><td colspan="5">评语：</td></tr>
</table>

学习情境 2-25

2.1.5.3 实施现场用电环节决策单

<table>
<tr><td>学习场</td><td colspan="5">金　工　实　训</td></tr>
<tr><td>学习情境 2</td><td colspan="3">台钻床临时用电</td><td>本环节学时</td><td>6 分钟</td></tr>
<tr><td>典型工作过程描述</td><td colspan="5">用电安全规范—分析用电技术要求—勘察现场用电环境—确定选用方案—实施现场用电</td></tr>
<tr><td colspan="6">计　划　对　比</td></tr>
<tr><td>序号</td><td>计划的可行性</td><td>计划的经济性</td><td>计划的可操作性</td><td>计划的实施难度</td><td>综合评价</td></tr>
<tr><td>方案 1</td><td></td><td></td><td></td><td></td><td></td></tr>
<tr><td>方案 2</td><td></td><td></td><td></td><td></td><td></td></tr>
<tr><td>方案 3</td><td></td><td></td><td></td><td></td><td></td></tr>
</table>

<table>
<tr><td rowspan="3">决策评价</td><td>班级</td><td></td><td>第（　　）组</td><td>组长签字</td><td></td></tr>
<tr><td>教师签字</td><td></td><td>日期</td><td colspan="2"></td></tr>
<tr><td colspan="5">评语：</td></tr>
</table>

2.1.5.4　实施现场用电环节实施单

<table>
<tr><td>学习场</td><td colspan="5">金　工　实　训</td></tr>
<tr><td>学习情境 2</td><td colspan="3">台钻床临时用电</td><td>本环节学时</td><td>40 分钟</td></tr>
<tr><td>典型工作过程描述</td><td colspan="5">用电安全规范—分析用电技术要求—勘察现场用电环境—确定选用方案—实施现场用电</td></tr>
<tr><td>序号</td><td>实 施 步 骤</td><td colspan="2">工作内容及方法</td><td colspan="2">注 意 事 项</td></tr>
<tr><td>1</td><td>作业前准备，落实现场安全措施</td><td colspan="2">检查工作人员的精神状态；办理工作许可手续；工作负责人进行现场安全交底，向工作人员指明工作任务、工作范围、带电部位及安全措施情况。</td><td colspan="2"></td></tr>
<tr><td>2</td><td>停电</td><td colspan="2">执行一人操作一人监护的制度，操作者戴上安全帽、戴好绝缘手套把电源开关拉下。</td><td colspan="2"></td></tr>
<tr><td>3</td><td>验电</td><td colspan="2"></td><td colspan="2"></td></tr>
<tr><td>4</td><td>安装设备、接线</td><td colspan="2"></td><td colspan="2"></td></tr>
<tr><td>5</td><td>送电</td><td colspan="2"></td><td colspan="2"></td></tr>
<tr><td>6</td><td>钻床的操作</td><td colspan="2"></td><td colspan="2"></td></tr>
<tr><td colspan="6">实施说明：</td></tr>
<tr><td rowspan="3">实施评价</td><td>班级</td><td></td><td>第（　　）组</td><td>组长签字</td><td></td></tr>
<tr><td>教师签字</td><td></td><td>日期</td><td colspan="2"></td></tr>
<tr><td colspan="5">评语：</td></tr>
</table>

2.1.5.5 实施现场用电环节检查单

学习场	金工实训				
学习情境 2	台钻床临时用电		本环节学时	6 分钟	
典型工作过程描述	用电安全规范—分析用电技术要求—勘察现场用电环境—确定选用方案—实施现场用电				
序号	检查项目	检 查 标 准		学生自查	教师检查
1	资讯环节	获得台钻床施工现场用电的有关信息的来源是否符合国家电力行业的标准。			
2	计划环节	制定台钻床施工现场用电步骤的各个方案。			
3	决策环节	对各种下料的方案衡量对比评分，获取最佳方案。			
4	实施环节	台钻床施工现场用电过程中各个环节按照标准执行，贯彻落实“四不伤害”。			
5	检查环节	各个环节逐一检查，是否有爱岗敬业的主题思想。			
检查评价	班级		第（ ）组	组长签字	
	教师签字		日期		
	评语：				

2.2　台钻床临时用电学习情境材料工具清单

学习场	金　工　实　训		
学习情境 2	台钻床临时用电	本环节学时	300 分钟
典型工作过程描述	用电安全规范—分析用电技术要求—勘察现场用电环境—确定选用方案—实施现场用电		

序号	名称	作用	数量	型号	使用者

班级		第（　　）组	组长签字	
教师签字		日期		

2.3 台钻床临时用电学习情境教学反馈单

<table>
<tr><td>学习场</td><td colspan="4">金 工 实 训</td></tr>
<tr><td>学习情境 2</td><td colspan="2">台钻床临时用电</td><td>本环节学时</td><td>300 分钟</td></tr>
<tr><td>典型工作过程描述</td><td colspan="4">用电安全规范—分析用电技术要求—勘察现场用电环境—确定选用方案—实施现场用电</td></tr>
<tr><td rowspan="6">调查项目</td><td>序号</td><td>调 查 内 容</td><td colspan="2">理 由 描 述</td></tr>
<tr><td></td><td></td><td colspan="2"></td></tr>
<tr><td></td><td></td><td colspan="2"></td></tr>
<tr><td></td><td></td><td colspan="2"></td></tr>
<tr><td></td><td></td><td colspan="2"></td></tr>
<tr><td></td><td></td><td colspan="2"></td></tr>
<tr><td colspan="5">您对本次课程教学的改进意见：</td></tr>
<tr><td>调查信息</td><td>被调查人姓名</td><td></td><td>调查日期</td><td></td></tr>
</table>

2.4　台钻床临时用电学习情境分组单

学习场	金　工　实　训					
学习情境 2	台钻床临时用电				本环节学时	300 分钟
典型工作过程描述	用电安全规范—分析用电技术要求—勘察现场用电环境—确定选用方案—实施现场用电					
分组情况	组别	组长	组　员			
分组说明：						
班级		教师签字		日期		

2.5 台钻床临时用电学习情境评价单

<table>
<tr><td>学习场</td><td colspan="5">金 工 实 训</td></tr>
<tr><td>学习情境 2</td><td colspan="2">台钻床临时用电</td><td colspan="2">本环节学时</td><td>20 分钟</td></tr>
<tr><td>典型工作过程描述</td><td colspan="5">用电安全规范—分析用电技术要求—勘察现场用电环境—确定选用方案—实施现场用电</td></tr>
<tr><td>评价项目</td><td colspan="2">评 价 子 项 目</td><td>学生自评
20%</td><td>组内评价
30%</td><td>教师评价
50%</td></tr>
<tr><td>用电安全规范典型工作环节</td><td colspan="2">1. 资讯 10 分；2. 计划 10 分；3. 决策 10 分；4. 实施 40 分；5. 检查 10 分；6. 工作态度 20 分。</td><td></td><td></td><td></td></tr>
<tr><td>分析用电技术要求典型工作环节</td><td colspan="2">1. 资讯 10 分；2. 计划 10 分；3. 决策 10 分；4. 实施 40 分；5. 检查 10 分；6. 工作态度 20 分。</td><td></td><td></td><td></td></tr>
<tr><td>勘察现场用电环境典型工作环节</td><td colspan="2">1. 资讯 10 分；2. 计划 10 分；3. 决策 10 分；4. 实施 40 分；5. 检查 10 分；6. 工作态度 20 分。</td><td></td><td></td><td></td></tr>
<tr><td>确定选用方案典型工作环节</td><td colspan="2">1. 资讯 10 分；2. 计划 10 分；3. 决策 10 分；4. 实施 40 分；5. 检查 10 分；6. 工作态度 20 分。</td><td></td><td></td><td></td></tr>
<tr><td>实施现场用电典型工作环节</td><td colspan="2">1. 资讯 10 分；2. 计划 10 分；3. 决策 10 分；4. 实施 40 分；5. 检查 10 分；6. 工作态度 20 分。</td><td></td><td></td><td></td></tr>
<tr><td>最终结果</td><td colspan="2">1. 资讯 10 分；2. 计划 10 分；3. 决策 10 分；4. 实施 40 分；5. 检查 10 分；6. 工作态度 20 分。</td><td></td><td></td><td></td></tr>
<tr><td rowspan="3">评价</td><td>班级</td><td></td><td>第（　　）组</td><td>组长签字</td><td></td></tr>
<tr><td>教师签字</td><td></td><td>日期</td><td colspan="2"></td></tr>
<tr><td colspan="5">评语：</td></tr>
</table>

学习情境 2-32

2.6　教师实施计划单

<table>
<tr><td>学习场</td><td colspan="6">金　工　实　训</td></tr>
<tr><td>学习情境 2</td><td colspan="3">台钻床临时用电</td><td colspan="2">学时</td><td>300 分钟</td></tr>
<tr><td>典型工作过程描述</td><td colspan="6">用电安全规范—分析用电技术要求—勘察现场用电环境—确定选用方案—实施现场用电</td></tr>
<tr><td>序号</td><td>工作与学习步骤</td><td>学时</td><td>使用工具</td><td>地点</td><td>方式</td><td>备注</td></tr>
<tr><td></td><td></td><td></td><td></td><td></td><td></td><td></td></tr>
<tr><td></td><td></td><td></td><td></td><td></td><td></td><td></td></tr>
<tr><td></td><td></td><td></td><td></td><td></td><td></td><td></td></tr>
<tr><td></td><td></td><td></td><td></td><td></td><td></td><td></td></tr>
<tr><td></td><td></td><td></td><td></td><td></td><td></td><td></td></tr>
<tr><td></td><td></td><td></td><td></td><td></td><td></td><td></td></tr>
<tr><td></td><td></td><td></td><td></td><td></td><td></td><td></td></tr>
<tr><td></td><td></td><td></td><td></td><td></td><td></td><td></td></tr>
<tr><td></td><td></td><td></td><td></td><td></td><td></td><td></td></tr>
<tr><td></td><td></td><td></td><td></td><td></td><td></td><td></td></tr>
<tr><td>班级</td><td></td><td>教师签字</td><td></td><td>日期</td><td colspan="2"></td></tr>
</table>

Learning scenario 2
Temporary electricity of bench drilling machine

2.0 Learning worksheet

<table>
<tr><td>Learning field</td><td colspan="6">Metalworking practical training</td></tr>
<tr><td>Learning scenario 2</td><td colspan="3">Temporary electricity of bench drilling machine</td><td colspan="2">Total teaching time</td><td>300 minutes</td></tr>
<tr><td>Description of the typical working processes</td><td colspan="6">Electricity safety code – analysing electricity technical requirements – on-site survey on the environment of electricity utilization – confirming the scheme selection – implementing on-site electricity utilization</td></tr>
<tr><td>Learning objectives</td><td colspan="6">1. Master the temporary electricity safety code on the construction site.
2. Master the electricity technical requirements for bench drilling machine on the construction site.
3. Master the selection of bench drilling machines on the construction site.
4. Establish professional awareness of electricity utilization safety and cultivate the dedication to work.</td></tr>
<tr><td>Task description</td><td colspan="6">A construction site needs to process and drill a hole in a 220kV cable cleat. Now, in accordance with the requirements, you have to connect the temporary power supply to the bench drilling machine. Please complete the task safely and effectively according to the on-site situation and requirements, you can refer to Figure 2.0.1.

Figure 2.0.1 On-site electricity utilization</td></tr>
<tr><td>Teaching schedule</td><td>Information: 30 minutes</td><td>Planning: 30 minutes</td><td>Decision: 30 minutes</td><td>Implementation: 160 minutes</td><td>Check: 30 minutes</td><td>Assessment: 20 minutes</td></tr>
<tr><td>Requirements for students</td><td colspan="6">Preview the practical training homework sheets before taking the class and use the Internet to find the learning materials related to our practical training. Wear the training uniform throughout the training process. Implement the “4 No-harm Principles” (do not harm yourself, do not harm others, do not get harmed by others, protect others from being harmed) and other safety requirements. Strictly follow the rules and regulations of the training room.</td></tr>
<tr><td>References</td><td colspan="6">Refer to Safety Code of Electric Power Industry — Electric Part of Power Plants and Transformer Substations (GB 26860—2011) and Technical Code for Safety of Temporary Electrification on Construction Site (JGJ 46—2005).</td></tr>
</table>

2.1 Teaching guidance

Learning field	Metalworking practical training	Learning scenario 2	Temporary electricity of bench drilling machine	Reference frame	*Technical Code for Safety of Temporary Electrification on Construction Site* (JGJ 46—2005)	
General working processes / Typical working processes	Information	Planning	Decision	Implementation	Check	Assessment
Electricity safety code	Check the *Technical Code Safety of for Temporary Electrification on Construction Site* (JGJ 46—2005) to obtain relevant information about the electricity safety code on the construction site, so as to complete the tasks safely and effectively.	Through group discussions among students, prepare an electricity safety code scheme and complete 2.1.1.2 Planning sheet for electricity safety code.	Take integrated evaluation based on the feasibility, economy, operability, and implementation difficulty of the plans, and then select the best one.	Check relevant information and complete 2.1.1.4 Implementation sheet for electricity safety code.	Conduct self-check by students and check by the teacher according to the check items and check criteria included in 2.1.1.5 Check sheet for electricity safety code.	Conduct self-assessment, peer assessment, and teacher assessment according to the points assigned to the evaluation items, and complete 2.5 Assessment sheet for the temporary electricity of bench drilling machine learning scenario.
Analysing electricity technical requirements	Check the *Technical Code Safety of for Temporary Electrification on Construction Site* (JGJ 46—2005) to obtain relevant information about the welding electricity technical requirements on the construction site, so as to complete the tasks safely and effec tively.	Through group discussions among students, prepare an electricity safety code scheme and complete 2.1.2.2 Planning sheet for analysing electricity technical requirements.	Take integrated evaluation based on the feasibility, economy, operability, and implementation difficulty of the plans, and then select the best one.	Check relevant information and complete 2.1.2.4 Implementation sheet for analysing electricity technical requirements.	Conduct self-check by students and check by the teacher according to the check items and check criteria included in 2.1.2.5 Check sheet for analysing electricity technical requirements.	Conduct self-assessment, peer assessment, and teacher assessment according to the points assigned to the evaluation items, and complete 2.5 Assessment sheet for the temporary electricity of bench drilling machine learning scenario.

Continue

Learning field	Metalworking practical training	Learning scenario 2	Temporary electricity of bench drilling machine	Reference frame	*Technical Code for Safety of Temporary Electrification on Construction Site* (JGJ 46—2005)	
General working processes / Typical working processes	Information	Planning	Decision	Implementation	Check	Assessment
On-site survey on the environment of electricity utilization	Check the *Technical Code Safety of for Temporary Electrification on Construction Site* (JGJ 46—2005) to obtain relevant information about the electricity safety code on the construction site, so as to complete the tasks safely and effectively.	Through group discussions among students, prepare an electricity safety code scheme and complete 2.1.3.2 Planning sheet for the on-site survey on the environment of electricity utilization.	Take integrated evaluation based on the feasibility, economy, operability, and implementation difficulty of the plans, and then select the best one.	Check relevant information and complete 2.1.3.4 Implementation sheet for the on-site survey on the environment of electricity utilization.	Conduct self-check by students and check by the teacher according to the check items and check criteria included in 2.1.3.5 Check sheet for the on-site survey on the environment of electricity utilization.	Conduct self-assessment, peer assessment, and teacher assessment according to the points assigned to the evaluation items, and complete 2.5 Assessment sheet for the temporary electricity of bench drilling machine learning scenario.
Confirming the scheme selection	Check the *Technical Code for Safety of Temporary Electrification on Construction Site* (JGJ 46—2005) to obtain relevant information about the electricity safety code on the construction site, so as to complete the tasks safely and effectively.	Through group discussions among students, prepare an electricity safety code scheme and complete 2.1.4.2 Planning sheet for confirming the scheme selection.	Take integrated evaluation based on the feasibility, economy, operability, and implementation difficulty of the plans, and then select the best one.	Check relevant information and complete 2.1.4.4 Implementation sheet for confirming the scheme selection.	Conduct self-check by students and check by the teacher according to the check items and check criteria included in 2.1.4.5 Check sheet for confirming the scheme selection.	Conduct self-assessment, peer assessment, and teacher assessment according to the points assigned to the evaluation items, and complete 2.5 Assessment sheet for the temporary electricity of bench drilling machine learning scenario.

Continue

Learning field	Metalworking practical training	Learning scenario 2	Temporary electricity of bench drilling machine	Reference frame	*Technical Code for Safety of Temporary Electrification on Construction Site* (JGJ 46—2005)	
General working processes / Typical working processes	Information	Planning	Decision	Implementation	Check	Assessment
Implementing on-site electricity utilization	Check the *Technical Code for Safety of Temporary Electrification on Construction Site* (JGJ 46—2005) to obtain relevant information about the welding electricity technical requirements on the construction site, so as to complete the tasks safely and effectively.	Through group discussions among students, prepare an electricity safety code scheme and complete 2.1.5.2 Planning sheet for implementing the on-site electricity utilization.	Take integrated evaluation based on the feasibility, economy, operability, and implementation difficulty of the plans, and then select the best one.	Check relevant information and complete 2.1.5.4 Implementation sheet for implementing the on-site electricity utilization.	Conduct self-check by students and check by the teacher according to the check items and check criteria included in 2.1.5.5 Check sheet for implementing the on-site electricity utilization.	Conduct self-assessment, peer assessment, and teacher assessment according to the points assigned to the evaluation items, and complete 2.5 Assessment sheet for the temporary electricity of bench drilling machine learning scenario.

2.1.1 Electricity safety code

2.1.1.1 Information sheet for electricity safety code

Learning field	Metalworking practical training		
Learning scenario 2	Temporary electricity of bench drilling machine	Session teaching time	6 minutes
Description of the typical working processes	Electricity safety code – analysing electricity technical requirements – on-site survey on the environment of electricity utilization – confirming the scheme selection – implementing on-site electricity utilization		
Information collection methods	Teacher's explanation, search on the Internet, or communication among students		
Information description	Review the materials provided by the instructor to obtain information about bench drilling machine electricity safety code, so as to complete the tasks safely and effectively, as indicated in Figure 2.1.1.  Figure 2.1.1 On-site survey		
Requirements for students	Preview the practical training homework sheets before taking the class and use the Internet to find the learning materials related to our practical training. Wear the training uniform throughout the training process. Implement the "4 No-harm Principles" (do not harm yourself, do not harm others, do not get harmed by others, protect others from being harmed) and other safety requirements. Strictly follow the rules and regulations of the training room.		
References	Refer to *Safety Code of Electric Power Industry — Electric Part of Power Plants and Transformer Substations* (GB 26860—2011) and *Technical Code for Safety of Temporary Electrification on Construction Site* (JGJ 46—2005).		

2.1.1.2　Planning sheet for electricity safety code

<table>
<tr><td colspan="2">Learning field</td><td colspan="5">Metalworking practical training</td></tr>
<tr><td colspan="2">Learning scenario 2</td><td colspan="2">Temporary electricity of bench drilling machine</td><td colspan="2">Session teaching time</td><td>6 minutes</td></tr>
<tr><td colspan="2">Description of the typical working processes</td><td colspan="5">Electricity safety code – analysing electricity technical requirements – on-site survey on the environment of electricity utilization – confirming the scheme selection – implementing on-site electricity utilization</td></tr>
<tr><td colspan="2">Planning methods</td><td colspan="5">Group discussion among classmates</td></tr>
<tr><td>Number</td><td colspan="2">Working steps</td><td colspan="4">Notice</td></tr>
<tr><td>1</td><td colspan="2">Conduct pre-class electricity safety education</td><td colspan="4"></td></tr>
<tr><td>2</td><td colspan="2">Watch the video about the on-site rescue procedures for electric shock</td><td colspan="4"></td></tr>
<tr><td>3</td><td colspan="2">Learn about the electricity safety knowledge on the construction site, relevant national industry standards, and the essence of relevant documents</td><td colspan="4"></td></tr>
<tr><td>4</td><td colspan="2">Discuss the learning benefits</td><td colspan="4"></td></tr>
<tr><td>5</td><td colspan="2">Analyse the handling of emergency accidents</td><td colspan="4"></td></tr>
<tr><td colspan="2" rowspan="3">Plan assessment</td><td>Class</td><td></td><td>Group ( )</td><td>Group leader's signature</td><td></td></tr>
<tr><td>Teacher's signature</td><td></td><td>Date</td><td colspan="2"></td></tr>
<tr><td colspan="5">Comments:</td></tr>
</table>

2.1.1.3 Decision sheet for electricity safety code

<table>
<tr><td>Learning field</td><td colspan="5">Metalworking practical training</td></tr>
<tr><td>Learning scenario 2</td><td colspan="3">Temporary electricity of bench drilling machine</td><td>Session teaching time</td><td>6 minutes</td></tr>
<tr><td>Description of the typical working processes</td><td colspan="5">Electricity safety code – analysing electricity technical requirements – on-site survey on the environment of electricity utilization – confirming the scheme selection – implementing on-site electricity utilization</td></tr>
<tr><td colspan="6">Plan comparison</td></tr>
<tr><td>Number</td><td>Feasibility</td><td>Economy</td><td>Operability</td><td>Implementation difficulty</td><td>Integrated evaluation</td></tr>
<tr><td>Plan 1</td><td></td><td></td><td></td><td></td><td></td></tr>
<tr><td>Plan 2</td><td></td><td></td><td></td><td></td><td></td></tr>
<tr><td>Plan 3</td><td></td><td></td><td></td><td></td><td></td></tr>
<tr><td rowspan="3">Decision assessment</td><td>Class</td><td></td><td>Group ( )</td><td>Group leader's signature</td><td></td></tr>
<tr><td>Teacher's signature</td><td></td><td>Date</td><td colspan="2"></td></tr>
<tr><td colspan="5">Comments:</td></tr>
</table>

2.1.1.4 Implementation sheet for electricity safety code

<table>
<tr><td colspan="2">Learning field</td><td colspan="4">Metalworking practical training</td></tr>
<tr><td colspan="2">Learning scenario 2</td><td colspan="2">Temporary electricity of bench drilling machine</td><td>Session teaching time</td><td>30 minutes</td></tr>
<tr><td colspan="2">Description of the typical working processes</td><td colspan="4">Electricity safety code – analysing electricity technical requirements – on-site survey on the environment of electricity utilization – confirming the scheme selection – implementing on-site electricity utilization</td></tr>
<tr><td>Number</td><td>Implementation steps</td><td colspan="3">Operations and methods</td><td>Notice</td></tr>
<tr><td>1</td><td>Conduct a pre-class meeting</td><td colspan="3">Conduct a pre-class meeting on electricity safety education.</td><td></td></tr>
<tr><td>2</td><td>Watch the video</td><td colspan="3">Watch the video about the on-site rescue procedures for electric shock.</td><td>Rescue skills for electric shock</td></tr>
<tr><td>3</td><td>Learn about the document standards</td><td colspan="3">Learn about the Safety Code of Electric Power Industry — Electric Part of Power Plants and Transformer Substations (GB 26860—2011), and Technical Code for Safety of Temporary Electrification on Construction Site (JGJ 46—2005). Search and study the diagram of the temporary electricity distribution box system, etc.</td><td></td></tr>
<tr><td>4</td><td>Discuss the learning benefits</td><td colspan="3"></td><td></td></tr>
<tr><td>5</td><td>Analyse the handling of emergency accidents</td><td colspan="3"></td><td></td></tr>
<tr><td colspan="6">Implementation instructions:</td></tr>
<tr><td rowspan="3">Implementation assessment</td><td>Class</td><td></td><td>Group ( )</td><td>Group leader's signature</td><td></td></tr>
<tr><td>Teacher's signature</td><td></td><td>Date</td><td colspan="2"></td></tr>
<tr><td colspan="5">Comments:</td></tr>
</table>

2.1.1.5 Check sheet for electricity safety code

<table>
<tr><td colspan="2">Learning field</td><td colspan="4">Metalworking practical training</td></tr>
<tr><td colspan="2">Learning scenario 2</td><td colspan="2">Temporary electricity of bench drilling machine</td><td>Session teaching time</td><td>6 minutes</td></tr>
<tr><td colspan="2">Description of the typical working processes</td><td colspan="4">Electricity safety code – analysing electricity technical requirements – on-site survey on the environment of electricity utilization – confirming the scheme selection – implementing on-site electricity utilization</td></tr>
<tr><td>Number</td><td>Check items</td><td>Check criteria</td><td>Self-check by students</td><td colspan="2">Check by the teacher</td></tr>
<tr><td>1</td><td>Information</td><td>Obtain the source of information about the on-site bench drilling machine electricity safety code.</td><td></td><td colspan="2"></td></tr>
<tr><td>2</td><td>Planning</td><td>Formulate relevant plans for bench drilling machine electricity safety code.</td><td></td><td colspan="2"></td></tr>
<tr><td>3</td><td>Decision</td><td>Compare scores of various bench drilling machine electricity safety code plans and select the best solution.</td><td></td><td colspan="2"></td></tr>
<tr><td>4</td><td>Implementation</td><td>Operate each step of the bench drilling machine electricity safety code in accordance with the standards, and strictly implement the “4 No-harm Principles”.</td><td></td><td colspan="2"></td></tr>
<tr><td>5</td><td>Check</td><td>Check each step carefully, and check whether the theme idea of dedication to work is followed.</td><td></td><td colspan="2"></td></tr>
</table>

<table>
<tr><td rowspan="3">Check assessment</td><td>Class</td><td></td><td>Group ( )</td><td>Group leader’s signature</td><td></td></tr>
<tr><td>Teacher’s signature</td><td></td><td>Date</td><td colspan="2"></td></tr>
<tr><td colspan="5">Comments:</td></tr>
</table>

2.1.2 Analysing electricity technical requirements

2.1.2.1 Information sheet of analysing electricity technical requirements

Learning field	Metalworking practical training		
Learning scenario 2	Temporary electricity of bench drilling machine	Session teaching time	6 minutes
Description of the typical working processes	Electricity safety code – analysing electricity technical requirements – on-site survey on the environment of electricity utilization – confirming the scheme selection – implementing on-site electricity utilization		
Information collection methods	Teacher's explanation, search on the Internet, or communication among students		
Information description	Review the materials provided by the instructor to obtain information about the on-site electricity safety code for bench drilling machine, so as to complete the tasks safely and effectively, as indicated in Figure 2.1.2.  Figure 2.1.2　Search for information		
Requirements for students	Preview the practical training homework sheets before taking the class and use the Internet to find the learning materials related to our practical training. Wear the training uniform throughout the training process. Implement the "4 No-harm Principles" (do not harm yourself, do not harm others, do not get harmed by others, protect others from being harmed) and other safety requirements. Strictly follow the rules and regulations of the training room.		
References	Refer to *Safety Code of Electric Power Industry — Electric Part of Power Plants and Transformer Substations*) (GB 26860—2011) and *Technical Code for Safety of Temporary Electrification on Construction Site* (JGJ 46—2005).		

2.1.2.2 Planning sheet for analysing electricity technical requirements

<table>
<tr><td colspan="2">Learning field</td><td colspan="4">Metalworking practical training</td></tr>
<tr><td colspan="2">Learning scenario 2</td><td colspan="2">Temporary electricity of bench drilling machine</td><td>Session teaching time</td><td>6 minutes</td></tr>
<tr><td colspan="2">Description of the typical working processes</td><td colspan="4">Electricity safety code – analysing electricity technical requirements – on-site survey on the environment of electricity utilization – confirming the scheme selection – implementing on-site electricity utilization</td></tr>
<tr><td colspan="2">Planning methods</td><td colspan="4">Group discussion among classmates</td></tr>
<tr><td>Number</td><td colspan="2">Working steps</td><td colspan="3">Notice</td></tr>
<tr><td>1</td><td colspan="2">Check the parameters of the bench drilling machine provided on site</td><td colspan="3">Check whether the bench drilling machine uses single-phase alternating current or three-phase alternating current; Check the power size of the mofor; Check whether the main spindle of the machine can change its rotational speed when the mofor uses the five-level variable speed pulley; Check whether the working table can move up and down over the round column.</td></tr>
<tr><td>2</td><td colspan="2">Analyse whether the normal power of the bench drilling machine fulfils the on-site hole drilling technical requirements</td><td colspan="3"></td></tr>
<tr><td>3</td><td colspan="2">Test the insulation performance of the bench drilling machine</td><td colspan="3"></td></tr>
<tr><td>4</td><td colspan="2">Analyse whether the working current of the distribution box air circuit breaker matches the working current of the bench drilling machine</td><td colspan="3"></td></tr>
<tr><td>5</td><td colspan="2">Analyse whether the power cord of the bench drilling machine is in line with the code and criteria</td><td colspan="3"></td></tr>
<tr><td>6</td><td colspan="2">Analyse whether the installation of earthing lead of the bench drilling machine is in line with the requirements</td><td colspan="3"></td></tr>
<tr><td colspan="2" rowspan="3">Plan assessment</td><td>Class</td><td></td><td>Group ( )</td><td>Group leader's signature</td></tr>
<tr><td>Teacher's signature</td><td></td><td>Date</td><td></td></tr>
<tr><td colspan="4">Comments:</td></tr>
</table>

2.1.2.3 Decision sheet for analysing electricity technical requirements

<table>
<tr><td>Learning field</td><td colspan="5">Metalworking practical training</td></tr>
<tr><td>Learning scenario 2</td><td colspan="3">Temporary electricity of bench drilling machine</td><td>Session teaching time</td><td>6 minutes</td></tr>
<tr><td>Description of the typical working processes</td><td colspan="5">Electricity safety code – analysing electricity technical requirements – on-site survey on the environment of electricity utilization – confirming the scheme selection – implementing on-site electricity utilization</td></tr>
<tr><td colspan="6">Plan comparison</td></tr>
<tr><td>Number</td><td>Feasibility</td><td>Economy</td><td>Operability</td><td>Implementation difficulty</td><td>Integrated evaluation</td></tr>
<tr><td>Plan 1</td><td></td><td></td><td></td><td></td><td></td></tr>
<tr><td>Plan 2</td><td></td><td></td><td></td><td></td><td></td></tr>
<tr><td>Plan 3</td><td></td><td></td><td></td><td></td><td></td></tr>
<tr><td rowspan="3">Decision assessment</td><td>Class</td><td></td><td>Group ( )</td><td>Group leader's signature</td><td></td></tr>
<tr><td>Teacher's signature</td><td></td><td>Date</td><td colspan="2"></td></tr>
<tr><td colspan="5">Comments:</td></tr>
</table>

2.1.2.4 Implementation sheet for analysing electricity technical requirements

<table>
<tr><td>Learning field</td><td colspan="5">Metalworking practical training</td></tr>
<tr><td>Learning scenario 2</td><td colspan="3">Temporary electricity of bench drilling machine</td><td>Session teaching time</td><td>30 minutes</td></tr>
<tr><td>Description of the typical working processes</td><td colspan="5">Electricity safety code – analysing electricity technical requirements – on-site survey on the environment of electricity utilization – confirming the scheme selection – implementing on-site electricity utilization</td></tr>
<tr><td>Number</td><td colspan="2">Implementation steps</td><td>Operations and methods</td><td colspan="2">Notice</td></tr>
<tr><td>1</td><td colspan="2">Check the parameters of the bench drilling machine provided on site</td><td></td><td colspan="2"></td></tr>
<tr><td>2</td><td colspan="2">Analyse whether the normal power of the bench drilling machine fulfils the on-site hole drilling technical requirements</td><td></td><td colspan="2"></td></tr>
<tr><td>3</td><td colspan="2">Test the insulation performance of the bench drilling machine</td><td></td><td colspan="2"></td></tr>
<tr><td>4</td><td colspan="2">Analyse whether the working current of the distribution box air circuit breaker matches the working current of the bench drilling machine</td><td></td><td colspan="2"></td></tr>
<tr><td>5</td><td colspan="2">Analyse whether the power cord of the bench drilling machine is in line with the code and criteria</td><td></td><td colspan="2"></td></tr>
<tr><td>6</td><td colspan="2">Analyse whether the installation of earthing lead of the bench drilling machine is in line with the requirements</td><td></td><td colspan="2"></td></tr>
<tr><td colspan="6">Implementation instructions:</td></tr>
<tr><td rowspan="3">Implementation assessment</td><td>Class</td><td></td><td>Group ( )</td><td>Group leader's signature</td><td></td></tr>
<tr><td>Teacher's signature</td><td></td><td>Date</td><td colspan="2"></td></tr>
<tr><td colspan="5">Comments:</td></tr>
</table>

2.1.2.5 Check sheet for analysing electricity technical requirements

<table>
<tr><td>Learning field</td><td colspan="5">Metalworking practical training</td></tr>
<tr><td>Learning scenario 2</td><td colspan="3">Temporary electricity of bench drilling machine</td><td>Session teaching time</td><td>6 minutes</td></tr>
<tr><td>Description of the typical working processes</td><td colspan="5">Electricity safety code – analysing electricity technical requirements – on-site survey on the environment of electricity utilization – confirming the scheme selection – implementing on-site electricity utilization</td></tr>
<tr><td>Number</td><td>Check items</td><td colspan="2">Check criteria</td><td>Self-check by students</td><td>Check by the teacher</td></tr>
<tr><td>1</td><td>Information</td><td colspan="2">Obtain the source of information about the on-site electricity safety code of the bench drilling machine, and identify whether it fulfils the national standard of the electric power industry.</td><td></td><td></td></tr>
<tr><td>2</td><td>Planning</td><td colspan="2">Formulate relevant plans for safe electricity utilization of the bench drilling machine on the construction site.</td><td></td><td></td></tr>
<tr><td>3</td><td>Decision</td><td colspan="2">Compare scores of various electricity utilization plans of the bench drilling machine and select the best solution.</td><td></td><td></td></tr>
<tr><td>4</td><td>Implementation</td><td colspan="2">Operate each step of the electricity utilization of the bench drilling machine in accordance with the standards, and strictly implement the “4 No-harm Principles”.</td><td></td><td></td></tr>
<tr><td>5</td><td>Check</td><td colspan="2">Check each step carefully, and check whether the theme idea of dedication to work is followed.</td><td></td><td></td></tr>
<tr><td rowspan="3">Check assessment</td><td>Class</td><td></td><td>Group ( )</td><td>Group leader’s signature</td><td></td></tr>
<tr><td>Teacher’s signature</td><td></td><td>Date</td><td colspan="2"></td></tr>
<tr><td colspan="5">Comments:</td></tr>
</table>

2.1.3 On-site survey on the environment of electricity utilization

2.1.3.1 Information sheet for on-site survey on the environment of electricity utilization

Learning field	Metalworking practical training		
Learning scenario 2	Temporary electricity of bench drilling machine	Session teaching time	6 minutes
Description of the typical working processes	Electricity safety code – analysing electricity technical requirements – on-site survey on the environment of electricity utilization – confirming the scheme selection – implementing on-site electricity utilization		
Information collection methods	Teacher's explanation, search on the Internet, or communication among students		
Information description	Review the materials provided by the teacher to obtain information about on-site electricity utilization environment of the bench drilling machine, so as to complete the tasks safely and effectively, as indicated in Figure 2.1.3.  Figure 2.1.3 On-site survey on the environment of electricity utilization		
Requirements for students	Preview the practical training homework sheets before taking the class and use the Internet to find the learning materials related to our practical training. Wear the training uniform throughout the training process. Implement the "4 No-harm Principles" (do not harm yourself, do not harm others, do not get harmed by others, protect others from being harmed) and other safety requirements. Strictly follow the rules and regulations of the training room.		
References	Refer to *Safety Code of Electric Power Industry — Electric Part of Power Plants and Transformer Substations* (GB 26860—2011) and *Technical Code for Safety of Temporary Electrification on Construction Site* (JGJ 46—2005).		

2.1.3.2 Planning sheet for on-site survey on the environment of electricity utilization

<table>
<tr><td>Learning field</td><td colspan="5">Metalworking practical training</td></tr>
<tr><td>Learning scenario 2</td><td colspan="3">Temporary electricity of bench drilling machine</td><td>Session teaching time</td><td>6 minutes</td></tr>
<tr><td>Description of the typical working processes</td><td colspan="5">Electricity safety code – analysing electricity technical requirements – on-site survey on the environment of electricity utilization – confirming the scheme selection – implementing on-site electricity utilization</td></tr>
<tr><td>Planning methods</td><td colspan="5">Group discussion among classmates</td></tr>
<tr><td>Number</td><td>Working steps</td><td colspan="4">Notice</td></tr>
<tr><td>1</td><td>Preparation before entering the construction site</td><td colspan="4"></td></tr>
<tr><td>2</td><td>Enter the construction site</td><td colspan="4"></td></tr>
<tr><td>3</td><td>On-site survey work</td><td colspan="4">Check whether adequate fire equipment is set up at the welding work area; Check whether the distance between the welding work area and the storage and usage of the oxygen cylinder, calcium carbide barrel, and acetylene generator meets the prescribed safety distance; Check whether the distance between the electricity source and the welding work area meets the prescribed safety distance; Check whether the power switch is equipped with the leak-proof electricity device; Check whether the surface of the welding work area is in line with the requirements; Check whether other protective measures are in place at the site.</td></tr>
<tr><td>4</td><td>Leave the construction site</td><td colspan="4"></td></tr>
<tr><td rowspan="3">Plan assessment</td><td>Class</td><td></td><td>Group ( )</td><td>Group leader's signature</td><td></td></tr>
<tr><td>Teacher's signature</td><td></td><td>Date</td><td colspan="2"></td></tr>
<tr><td colspan="5">Comments:</td></tr>
</table>

2.1.3.3 Decision sheet for on-site survey on the environment of electricity utilization

<table>
<tr><td>Learning field</td><td colspan="5">Metalworking practical training</td></tr>
<tr><td>Learning scenario 2</td><td colspan="3">Temporary electricity of bench drilling machine</td><td>Session teaching time</td><td>6 minutes</td></tr>
<tr><td>Description of the typical working processes</td><td colspan="5">Electricity safety code – analysing electricity technical requirements – on-site survey on the environment of electricity utilization – confirming the scheme selection – implementing on-site electricity utilization</td></tr>
<tr><td colspan="6">Plan comparison</td></tr>
<tr><td>Number</td><td>Feasibility</td><td>Economy</td><td>Operability</td><td>Implementation difficulty</td><td>Integrated evaluation</td></tr>
<tr><td>Plan 1</td><td></td><td></td><td></td><td></td><td></td></tr>
<tr><td>Plan 2</td><td></td><td></td><td></td><td></td><td></td></tr>
<tr><td>Plan 3</td><td></td><td></td><td></td><td></td><td></td></tr>
<tr><td rowspan="3">Decision assessment</td><td>Class</td><td></td><td>Group ( )</td><td>Group leader's signature</td><td></td></tr>
<tr><td>Teacher's signature</td><td></td><td>Date</td><td colspan="2"></td></tr>
<tr><td colspan="5">Comments:</td></tr>
</table>

2.1.3.4 Implementation sheet for on-site survey on the environment of electricity utilization

<table>
<tr><td>Learning field</td><td colspan="5">Metalworking practical training</td></tr>
<tr><td>Learning scenario 2</td><td colspan="2">Temporary electricity of bench drilling machine</td><td colspan="2">Session teaching time</td><td>30 minutes</td></tr>
<tr><td>Description of the typical working processes</td><td colspan="5">Electricity safety code – analysing electricity technical requirements – on-site survey on the environment of electricity utilization – confirming the scheme selection – implementing on-site electricity utilization</td></tr>
<tr><td>Number</td><td>Implementation steps</td><td colspan="2">Operations and methods</td><td colspan="2">Notice</td></tr>
<tr><td>1</td><td>Preparation before entering the construction site</td><td colspan="2">Check whether you are neatly dressed and wearing a helmet and other personal protective equipment correctly.</td><td colspan="2"></td></tr>
<tr><td>2</td><td>Enter the construction site</td><td colspan="2"></td><td colspan="2"></td></tr>
<tr><td>3</td><td>On-site survey work</td><td colspan="2">Check whether adequate fire equipment is set up at the welding work area; Check whether the distance between the welding work area and the storage and usage of the oxygen cylinder, calcium carbide barrel, and acetylene generator meets the prescribed safety distance; Check whether the distance between the electricity source and the welding work area meets the prescribed safety distance; Check whether the power switch is equipped with the leak-proof electricity device; Check whether the ground of the welding work area is in line with the requirements; Check whether other protective measures are in place at the site.</td><td colspan="2"></td></tr>
<tr><td>4</td><td>Leave the construction site</td><td colspan="2"></td><td colspan="2"></td></tr>
<tr><td colspan="6">Implementation instructions:</td></tr>
<tr><td rowspan="3">Implementation assessment</td><td>Class</td><td></td><td>Group ( )</td><td>Group leader's signature</td><td></td></tr>
<tr><td>Teacher's signature</td><td></td><td>Date</td><td colspan="2"></td></tr>
<tr><td colspan="5">Comments:</td></tr>
</table>

2.1.3.5 Check sheet for on-site survey on the environment of electricity utilization

<table>
<tr><td colspan="2">Learning field</td><td colspan="4">Metalworking practical training</td></tr>
<tr><td colspan="2">Learning scenario 2</td><td colspan="2">Temporary electricity of bench drilling machine</td><td>Session teaching time</td><td>6 minutes</td></tr>
<tr><td colspan="2">Description of the typical working processes</td><td colspan="4">Electricity safety code – analysing electricity technical requirements – on-site survey on the environment of electricity utilization – confirming the scheme selection – implementing on-site electricity utilization</td></tr>
<tr><td>Number</td><td>Check items</td><td colspan="2">Check criteria</td><td>Self-check by students</td><td>Check by the teacher</td></tr>
<tr><td>1</td><td>Information</td><td colspan="2">Obtain the source of information about the on-site survey on the electricity utilization environment of the bench drilling machine, and check whether it fulfills the national industry standards.</td><td></td><td></td></tr>
<tr><td>2</td><td>Planning</td><td colspan="2">Formulate relevant plans for on-site survey on the electricity utilization environment of the bench drilling machine.</td><td></td><td></td></tr>
<tr><td>3</td><td>Decision</td><td colspan="2">Compare scores of various on-site survey plans on the electricity utilization environment of the bench drilling machine and select the best solution.</td><td></td><td></td></tr>
<tr><td>4</td><td>Implementation</td><td colspan="2">Operate each step of the on-site survey on the electricity utilization environment of the bench drilling machine in accordance with the standards, and strictly implement the “4 No-harm Principles”.</td><td></td><td></td></tr>
<tr><td>5</td><td>Check</td><td colspan="2">Check each step carefully, and check whether the theme idea of dedication to work is followed.</td><td></td><td></td></tr>
<tr><td colspan="2" rowspan="3">Check assessment</td><td>Class</td><td></td><td>Group ( )</td><td>Group leader’s signature</td></tr>
<tr><td>Teacher’s signature</td><td></td><td>Date</td><td></td></tr>
<tr><td colspan="4">Comments:</td></tr>
</table>

2.1.4 Confirming the scheme selection

2.1.4.1 Information sheet for confirming the scheme selection

<table>
<tr><td>Learning field</td><td colspan="3">Metalworking practical training</td></tr>
<tr><td>Learning scenario 2</td><td>Temporary electricity of bench drilling machine</td><td>Session teaching time</td><td>6 minutes</td></tr>
<tr><td>Description of the typical working processes</td><td colspan="3">Electricity safety code – analysing electricity technical requirements – on-site survey on the environment of electricity utilization – confirming the scheme selection – implementing on-site electricity utilization</td></tr>
<tr><td>Information collection methods</td><td colspan="3">Teacher’s explanation, search on the Internet, or communication among students</td></tr>
<tr><td>Information description</td><td colspan="3">Review the materials provided by the teacher to obtain information about the on-site selection of bench drilling machine, so as to complete the tasks safely and effectively, as indicated in Figure 2.1.4.

Figure 2.1.4　Confirming the scheme selection</td></tr>
<tr><td>Requirements for students</td><td colspan="3">Preview the practical training homework sheets before taking the class and use the Internet to find the learning materials related to our practical training. Wear the training uniform throughout the training process. Implement the “4 No-harm Principles” (do not harm yourself, do not harm others, do not get harmed by others, protect others from being harmed) and other safety requirements. Strictly follow the rules and regulations of the training room.</td></tr>
<tr><td>References</td><td colspan="3">Refer to Safety Code of Electric Power Industry — Electric Part of Power Plants and Transformer Substations (GB 26860—2011) and Technical Code for Safety of Temporary Electrification on Construction Site (JGJ 46—2005).</td></tr>
</table>

2.1.4.2 Planning sheet for confirming the scheme selection

<table>
<tr><td>Learning field</td><td colspan="5">Metalworking practical training</td></tr>
<tr><td>Learning scenario 2</td><td colspan="3">Temporary electricity of bench drilling machine</td><td>Session teaching time</td><td>6 minutes</td></tr>
<tr><td>Description of the typical working processes</td><td colspan="5">Electricity safety code – analysing electricity technical requirements – on-site survey on the environment of electricity utilization – confirming the scheme selection – implementing on-site electricity utilization</td></tr>
<tr><td>Planning methods</td><td colspan="5">Group discussion among classmates</td></tr>
<tr><td>Number</td><td>Working steps</td><td colspan="4">Notice</td></tr>
<tr><td>1</td><td>Inquire about the related standards</td><td colspan="4"></td></tr>
<tr><td>2</td><td>Compose the scheme</td><td colspan="4"></td></tr>
<tr><td>3</td><td>Scheme submission</td><td colspan="4"></td></tr>
<tr><td>4</td><td>Group discussion</td><td colspan="4"></td></tr>
<tr><td>5</td><td>Confirm the scheme</td><td colspan="4"></td></tr>
<tr><td rowspan="3">Plan assessment</td><td>Class</td><td></td><td>Group ( )</td><td>Group leader's signature</td><td></td></tr>
<tr><td>Teacher's signature</td><td></td><td>Date</td><td colspan="2"></td></tr>
<tr><td colspan="5">Comments:</td></tr>
</table>

2.1.4.3 Decision sheet for confirming the scheme selection

<table>
<tr><td>Learning field</td><td colspan="5">Metalworking practical training</td></tr>
<tr><td>Learning scenario 2</td><td colspan="3">Temporary electricity of bench drilling machine</td><td>Session teaching time</td><td>6 minutes</td></tr>
<tr><td>Description of the typical working processes</td><td colspan="5">Electricity safety code – analysing electricity technical requirements – on-site survey on the environment of electricity utilization – confirming the scheme selection – implementing on-site electricity utilization</td></tr>
<tr><td colspan="6">Plan comparison</td></tr>
<tr><td>Number</td><td>Feasibility</td><td>Economy</td><td>Operability</td><td>Implementation difficulty</td><td>Integrated evaluation</td></tr>
<tr><td>Plan 1</td><td></td><td></td><td></td><td></td><td></td></tr>
<tr><td>Plan 2</td><td></td><td></td><td></td><td></td><td></td></tr>
<tr><td>Plan 3</td><td></td><td></td><td></td><td></td><td></td></tr>
<tr><td rowspan="3">Decision assessment</td><td>Class</td><td></td><td>Group ( )</td><td>Group leader's signature</td><td></td></tr>
<tr><td>Teacher's signature</td><td></td><td>Date</td><td colspan="2"></td></tr>
<tr><td colspan="5">Comments:</td></tr>
</table>

2.1.4.4 Implementation sheet for confirming the scheme selection

Learning field	Metalworking practical training		
Learning scenario 2	Temporary electricity of bench drilling machine	Session teaching time	30 minutes
Description of the typical working processes	Electricity safety code – analysing electricity technical requirements – on-site survey on the environment of electricity utilization – confirming the scheme selection – implementing on-site electricity utilization		

Number	Implementation steps	Operations and methods	Notice
1	Inquire about the related standards		
2	Compose the scheme		
3	Scheme submission		
4	Group discussion		
5	Confirm the scheme		

Implementation instructions:

Implementation assessment	Class		Group ( )	Group leader's signature	
	Teacher's signature		Date		
	Comments:				

2.1.4.5 Check sheet for confirming the scheme selection

<table>
<tr><td colspan="2">Learning field</td><td colspan="4">Metalworking practical training</td></tr>
<tr><td colspan="2">Learning scenario 2</td><td colspan="2">Temporary electricity of bench drilling machine</td><td>Session teaching time</td><td>6 minutes</td></tr>
<tr><td colspan="2">Description of the typical working processes</td><td colspan="4">Electricity safety code – analysing electricity technical requirements – on-site survey on the environment of electricity utilization – confirming the scheme selection – implementing on-site electricity utilization</td></tr>
<tr><td>Number</td><td>Check items</td><td>Check criteria</td><td>Self-check by students</td><td colspan="2">Check by the teacher</td></tr>
<tr><td>1</td><td>Information</td><td>Obtain the source of information about the on-site bench drilling machine selection, and identify whether it fulfils the national standard of the electric power industry.</td><td></td><td colspan="2"></td></tr>
<tr><td>2</td><td>Planning</td><td>Formulate relevant plans for bench drilling machine selection on the construction site.</td><td></td><td colspan="2"></td></tr>
<tr><td>3</td><td>Decision</td><td>Compare scores of various bench drilling machine selection schemes on the construction site and select the best solution.</td><td></td><td colspan="2"></td></tr>
<tr><td>4</td><td>Implementation</td><td>Operate each step of on-site bench drilling machine selection in accordance with the standards, and strictly implement the “4 No-harm Principles”.</td><td></td><td colspan="2"></td></tr>
<tr><td>5</td><td>Check</td><td>Check each step carefully, and check whether the theme idea of dedication to work is followed.</td><td></td><td colspan="2"></td></tr>
<tr><td colspan="2" rowspan="3">Check assessment</td><td>Class</td><td></td><td>Group ( )</td><td>Group leader’s signature</td></tr>
<tr><td>Teacher’s signature</td><td></td><td>Date</td><td></td></tr>
<tr><td colspan="4">Comments:</td></tr>
</table>

2.1.5 Implementing on-site electricity utilization

2.1.5.1 Information sheet for implementing on-site electricity utilization

<table>
<tr><td>Learning field</td><td colspan="3">Metalworking practical training</td></tr>
<tr><td>Learning scenario 2</td><td>Temporary electricity of bench drilling machine</td><td>Session teaching time</td><td>6 minutes</td></tr>
<tr><td>Description of the typical working processes</td><td colspan="3">Electricity safety code – analysing electricity technical requirements – on-site survey on the environment of electricity utilization – confirming the scheme selection – implementing on-site electricity utilization</td></tr>
<tr><td>Information collection methods</td><td colspan="3">Teacher's explanation, search on the Internet, or communication among students</td></tr>
<tr><td>Information description</td><td colspan="3">Review the materials provided by the instructor to obtain information about the on-site electricity utilization for bench drilling machine, so as to complete the tasks safely and effectively, as indicated in Figure 2.1.5.

Figure 2.1.5 On-site electricity utilization</td></tr>
<tr><td>Requirements for students</td><td colspan="3">Preview the practical training homework sheets before taking the class and use the Internet to find the learning materials related to our practical training. Wear the training uniform throughout the training process. Implement the "4 No-harm Principles" (do not harm yourself, do not harm others, do not get harmed by others, protect others from being harmed) and other safety requirements. Strictly follow the rules and regulations of the training room.</td></tr>
<tr><td>References</td><td colspan="3">Refer to Safety Code of Electric Power Industry — Electric Part of Power Plants and Transformer Substations (GB 26860—2011) and Technical Code for Safety of Temporary Electrification on Construction Site (JGJ 46—2005).</td></tr>
</table>

2.1.5.2 Planning sheet for implementing on-site electricity utilization

<table>
<tr><td>Learning field</td><td colspan="3">Metalworking practical training</td></tr>
<tr><td>Learning scenario 2</td><td>Temporary electricity of bench drilling machine</td><td>Session teaching time</td><td>6 minutes</td></tr>
<tr><td>Description of the typical working processes</td><td colspan="3">Electricity safety code – analysing electricity technical requirements – on-site survey on the environment of electricity utilization – confirming the scheme selection – implementing on-site electricity utilization</td></tr>
<tr><td>Planning methods</td><td colspan="3">Group discussion among classmates</td></tr>
</table>

<table>
<tr><td>Number</td><td>Working steps</td><td>Operations</td><td>Notice</td></tr>
<tr><td>1</td><td>Preparation before the operation, implement on-site safety measures</td><td>Check the spirits of the staff; Carry out work permit procedures; The person in charge should conduct a handover to the staff on on-site safety and specify the task, the work area, the parts with electricity, and situations of safety measures.</td><td></td></tr>
<tr><td>2</td><td>Electricity cut</td><td></td><td></td></tr>
<tr><td>3</td><td>Electricity check</td><td></td><td></td></tr>
<tr><td>4</td><td>Device installation, wiring</td><td></td><td></td></tr>
<tr><td>5</td><td>Electricity transmission</td><td></td><td></td></tr>
<tr><td>6</td><td>Operation of bench drilling</td><td></td><td></td></tr>
</table>

<table>
<tr><td rowspan="3">Plan assessment</td><td>Class</td><td></td><td>Group ( )</td><td>Group leader's signature</td><td></td></tr>
<tr><td>Teacher's signature</td><td></td><td>Date</td><td colspan="2"></td></tr>
<tr><td colspan="5">Comments:</td></tr>
</table>

2.1.5.3 Decision sheet for implementing on-site electricity utilization

<table>
<tr><td>Learning field</td><td colspan="5">Metalworking practical training</td></tr>
<tr><td>Learning scenario 2</td><td colspan="3">Temporary electricity of bench drilling machine</td><td>Session teaching time</td><td>6 minutes</td></tr>
<tr><td>Description of the typical working processes</td><td colspan="5">Electricity safety code – analysing electricity technical requirements – on-site survey on the environment of electricity utilization – confirming the scheme selection – implementing on-site electricity utilization</td></tr>
<tr><td colspan="6">Plan comparison</td></tr>
<tr><td>Number</td><td>Feasibility</td><td>Economy</td><td>Operability</td><td>Implementation difficulty</td><td>Integrated evaluation</td></tr>
<tr><td>Plan 1</td><td></td><td></td><td></td><td></td><td></td></tr>
<tr><td>Plan 2</td><td></td><td></td><td></td><td></td><td></td></tr>
<tr><td>Plan 3</td><td></td><td></td><td></td><td></td><td></td></tr>
<tr><td rowspan="3">Decision assessment</td><td>Class</td><td></td><td>Group ( )</td><td>Group leader's signature</td><td></td></tr>
<tr><td>Teacher's signature</td><td></td><td>Date</td><td></td><td></td></tr>
<tr><td colspan="5">Comments:</td></tr>
</table>

2.1.5.4 Implementation sheet for implementing on-site electricity utilization

<table>
<tr><td>Learning field</td><td colspan="4">Metalworking practical training</td></tr>
<tr><td>Learning scenario 2</td><td colspan="2">Temporary electricity of bench drilling machine</td><td>Session teaching time</td><td>40 minutes</td></tr>
<tr><td>Description of the typical working processes</td><td colspan="4">Electricity safety code – analysing electricity technical requirements – on-site survey on the environment of electricity utilization – confirming the scheme selection – implementing on-site electricity utilization</td></tr>
<tr><td>Number</td><td>Implementation steps</td><td colspan="2">Operations and methods</td><td>Notice</td></tr>
<tr><td>1</td><td>Preparation before the operation, implement on-site safety measures</td><td colspan="2">Check the spirits of the staff; Carry out work permit procedures; The person in charge should conduct a handover to the staff on on-site safety and specify the task, the work area, the parts with electricity, and situations of safety measures.</td><td></td></tr>
<tr><td>2</td><td>Electricity cut</td><td colspan="2">Implement the rule of “one person operating while one person supervising”. The operator should wear a helmet and insulated gloves and pull down the power switch.</td><td></td></tr>
<tr><td>3</td><td>Electricity check</td><td colspan="2"></td><td></td></tr>
<tr><td>4</td><td>Device installation, wiring</td><td colspan="2"></td><td></td></tr>
<tr><td>5</td><td>Electricity transmission</td><td colspan="2"></td><td></td></tr>
<tr><td>6</td><td>Operation of bench drilling</td><td colspan="2"></td><td></td></tr>
<tr><td colspan="5">Implementation instructions:</td></tr>
</table>

<table>
<tr><td rowspan="3">Implementation assessment</td><td>Class</td><td></td><td>Group ( )</td><td>Group leader’s signature</td><td></td></tr>
<tr><td>Teacher’s signature</td><td></td><td>Date</td><td colspan="2"></td></tr>
<tr><td colspan="5">Comments:</td></tr>
</table>

2.1.5.5 Check sheet for implementing on-site electricity utilization

Learning field	Metalworking practical training		
Learning scenario 2	Temporary electricity of bench drilling machine	Session teaching time	6 minutes
Description of the typical working processes	Electricity safety code – analysing electricity technical requirements – on-site survey on the environment of electricity utilization – confirming the scheme selection – implementing on-site electricity utilization		

Number	Check items	Check criteria	Self-check by students	Check by the teacher
1	Information	Obtain the source of information about the on-site electricity utilization of the bench drilling machine, and check whether it fulfils the national industry standards.		
2	Planning	Formulate relevant plans for electricity utilization of the bench drilling machine on the construction site.		
3	Decision	Compare scores of various cutting stock plans and select the best solution.		
4	Implementation	Operate each step of the on-site electricity utilization of the bench drilling machine in accordance with the standards, and strictly implement the "4 No-harm Principles".		
5	Check	Check each step carefully, and check whether the theme idea of dedication to work is followed.		

Check assessment	Class		Group ( )	Group leader's signature	
	Teacher's signature		Date		
	Comments:				

2.2　List of materials and tools for the temporary electricity of bench drilling machine learning scenario

Learning field	Metalworking practical training				
Learning scenario 2	Temporary electricity of bench drilling machine		Session teaching time		300 minutes
Description of the typical working processes	Electricity safety code – analysing electricity technical requirements – on-site survey on the environment of electricity utilization – confirming the scheme selection – implementing on-site electricity utilization				
Number	Name	Function	Quantity	Model	User
Class		Group ( )	Group leader's signature		
Teacher's signature		Date			

2.3 Teaching Feedback Sheet for the temporary electricity of bench drilling machine

<table>
<tr><td>Learning field</td><td colspan="4">Metalworking practical training</td></tr>
<tr><td>Learning scenario 2</td><td colspan="2">Temporary electricity of bench drilling machine</td><td>Session teaching time</td><td>300 minutes</td></tr>
<tr><td>Description of the typical working processes</td><td colspan="4">Electricity safety code – analysing electricity technical requirements – on-site survey on the environment of electricity utilization – confirming the scheme selection – implementing on-site electricity utilization</td></tr>
<tr><td rowspan="7">Survey items</td><td>Number</td><td colspan="2">Survey contents</td><td>Description of the reasons</td></tr>
<tr><td></td><td colspan="2"></td><td></td></tr>
<tr><td></td><td colspan="2"></td><td></td></tr>
<tr><td></td><td colspan="2"></td><td></td></tr>
<tr><td></td><td colspan="2"></td><td></td></tr>
<tr><td></td><td colspan="2"></td><td></td></tr>
<tr><td></td><td colspan="2"></td><td></td></tr>
<tr><td colspan="5">Your feedback on how to improve the teaching of this course:</td></tr>
<tr><td>Survey information</td><td>Respondent’s name</td><td></td><td>Survey date</td><td></td></tr>
</table>

2.4 Grouping sheet for the temporary electricity of bench drilling machine learning scenario

Learning field	Metalworking practical training					
Learning scenario 2	Temporary electricity of bench drilling machine		Session teaching time		300 minutes	
Description of the typical working processes	Electricity safety code – analysing electricity technical requirements – on-site survey on the environment of electricity utilization – confirming the scheme selection – implementing on-site electricity utilization					
Grouping	Group	Group leader	Group members			
Grouping instructions						
Class		Teacher's signature			Date	

2.5 Assessment sheet for the temporary electricity of bench drilling machine learning scenario

Learning field	Metalworking practical training			
Learning scenario 2	Temporary electricity of bench drilling machine	Session teaching time	20 minutes	
Description of the typical working processes	Electricity safety code – analysing electricity technical requirements – on-site survey on the environment of electricity utilization – confirming the scheme selection – implementing on-site electricity utilization			
Assessment items	Assessment subitems	Self-assessment by students 20%	Intra-group assessment 30%	Assessment by the teacher 50%
Electricity safety code	1. Information: 10 points; 2. Planning: 10 points; 3. Decision: 10 points; 4. Implementation: 40 points; 5. Check: 10 points; 6. Work attitude: 20 points.			
Analysing electricity technical requirements	1. Information: 10 points; 2. Planning: 10 points; 3. Decision: 10 points; 4. Implementation: 40 points; 5. Check: 10 points; 6. Work attitude: 20 points.			
On-site survey on the environment of electricity utilization	1. Information: 10 points; 2. Planning: 10 points; 3. Decision: 10 points; 4. Implementation: 40 points; 5. Check: 10 points; 6. Work attitude: 20 points.			
Confirming the scheme selection	1. Information: 10 points; 2. Planning: 10 points; 3. Decision: 10 points; 4. Implementation: 40 points; 5. Check: 10 points; 6. Work attitude: 20 points.			
Implementing on-site electricity utilization	1. Information: 10 points; 2. Planning: 10 points; 3. Decision: 10 points; 4. Implementation: 40 points; 5. Check: 10 points; 6. Work attitude: 20 points.			

Continue

<table>
<tr><td>Assessment items</td><td colspan="2">Assessment subitems</td><td>Self-assessment by students 20%</td><td>Intra-group assessment 30%</td><td>Assessment by the teacher 50%</td></tr>
<tr><td>Final results</td><td colspan="2">1. Information: 10 points;
2. Planning: 10 points;
3. Decision: 10 points;
4. Implementation: 40 points;
5. Check: 10 points;
6. Work attitude: 20 points.</td><td></td><td></td><td></td></tr>
<tr><td rowspan="3">Assessment</td><td>Class</td><td></td><td>Group ( )</td><td>Group leader's signature</td><td></td></tr>
<tr><td>Teacher's signature</td><td></td><td>Date</td><td colspan="2"></td></tr>
<tr><td colspan="5">Comments:</td></tr>
</table>

2.6 Teacher's implementation planning sheet

Learning field	Metalworking practical training					
Learning scenario 2	Temporary electricity of bench drilling machine			Session teaching time		300 minutes
Description of the typical working processes	Electricity safety code – analysing electricity technical requirements – on-site survey on the environment of electricity utilization – confirming the scheme selection – implementing on-site electricity utilization					
Number	Working and learning steps	Teaching time	Tools	Locations	Methods	Notes
Class		Teacher's signature		Date		

学习情境 3

变压器低压套管线夹 M10 六角螺母的加工

3.0 学习性工作任务单

<table>
<tr><td>学习场</td><td colspan="6">金 工 实 训</td></tr>
<tr><td>学习情境 3</td><td colspan="3">变压器低压套管线夹 M10 六角螺母的加工</td><td colspan="2">总学时</td><td>540 分钟</td></tr>
<tr><td>典型工作过程描述</td><td colspan="6">下料—锯割—锉端面—划端面正六边形—锉正六棱面—钻孔—攻螺纹—倒角</td></tr>
<tr><td>学习目标</td><td colspan="6">1. 能正确使用常用工具、量具和独立完成简单零件加工。
2. 能够独立完成含有划线、锯割、锉削的工件加工。
3. 能安全规范操作台钻。</td></tr>
<tr><td>任务描述</td><td colspan="6">把一块 ϕ25mm 的圆钢经过下料—锉端面 A—锯割—锉端面 B—划端面正六边形—锉正六棱面—钻孔—攻螺纹—倒角—修整，得到一个 M10 六角螺母，其中 $m=8.4$mm，$S=16$mm，$e=17.77$mm，如图 3.0.1 所示。
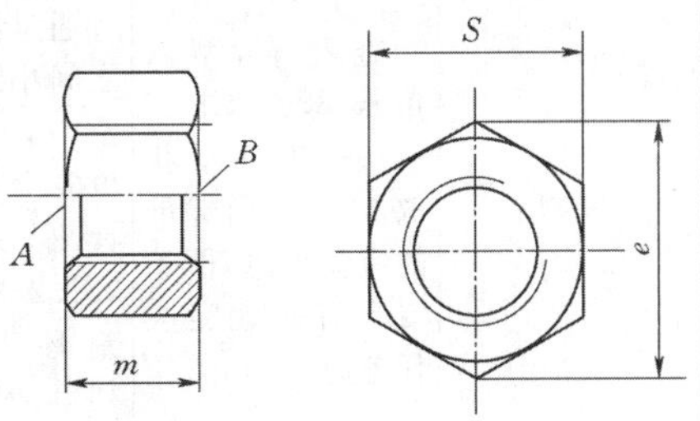

(a) 棱面 (b) 端面
图 3.0.1 六角螺母</td></tr>
<tr><td rowspan="2">学时安排</td><td>资讯</td><td>计划</td><td>决策</td><td>实施</td><td>检查</td><td>评价</td></tr>
<tr><td>50 分钟</td><td>50 分钟</td><td>50 分钟</td><td>320 分钟</td><td>50 分钟</td><td>20 分钟</td></tr>
<tr><td>对学生的要求</td><td colspan="6">课前充分预习实训作业单，充分利用网络查找有关实训的学习资料，实训过程中穿着实训服装，贯彻落实自己不伤害自己、自己不伤害他人、自己不被他人伤害、保护他人不被伤害的“四不伤害”安全要求，注意严格遵守实训室的各项规章制度。</td></tr>
<tr><td>参考资料</td><td colspan="6">参考《施工现场临时用电安全技术规范》(JGJ 46—2005) 和《1 型六角螺母》(GB/T 6170—2015)。</td></tr>
</table>

3.1　教学引导文

学习场	金工实训	学习情境 3	变压器低压套管线夹M10六角螺母的加工	参照系	《1 型六角螺母》（GB/T 6170—2015）	
普适性工作过程 典型工作过程	资讯	计划	决策	实施	检查	评价
下料	查看《1 型六角螺母》（GB/T 6170—2015），获取 M10 六角螺母的有关信息，便于安全有效地完成任务。	通过同学之间小组讨论，拟订用电安全规范计划，完成 3.1.1.2 下料环节计划单。	计划的可行性、经济性、可操作性和实施难度几方面综合分析，选取最佳方案。	查找其他有关资料，完成 3.1.1.4 下料环节实施单。	按照 3.1.1.5 下料环节检查单的检查项目和标准，进行学生自查和教师检查。	根据评价项目配置分值，进行自评、互评和教师评价，完成 3.5 变压器低压套管线夹 M10 六角螺母的加工学习情境评价单。
锯割	查看《1 型六角螺母》（GB/T 6170—2015），获取 M10 六角螺母的有关信息，便于安全有效地完成任务。	通过同学之间小组讨论，拟订用电安全规范计划，完成 3.1.2.2 锯割环节计划单。	计划的可行性、经济性、可操作性和实施难度几方面综合分析，选取最佳方案。	查找有关资料，完成 3.1.2.4 锯割环节实施单。	按照 3.1.2.5 锯割环节检查单的检查项目和标准，进行学生自查和教师检查。	根据评价项目配置分值，进行自评、互评和教师评价，完成 3.5 变压器低压套管线夹 M10 六角螺母的加工学习情境评价单。
锉端面	查看《1 型六角螺母》（GB/T 6170—2015），获取 M10 六角螺母的有关信息，便于安全有效地完成任务。	通过同学之间小组讨论，拟订用电安全规范计划，完成 3.1.3.2 锉端面环节计划单。	计划的可行性、经济性、可操作性和实施难度几方面综合分析，选取最佳方案。	查找有关资料，完成 3.1.3.4 锉端面环节实施单。	按照 3.1.3.5 锉端面环节检查单的检查项目和标准，进行学生自查和教师检查。	根据评价项目配置分值，进行自评、互评和教师评价，完成 3.5 变压器低压套管线夹 M10 六角螺母的加工学习情境评价单。

续表

学习场	金工实训	学习情境 3	变压器低压套管线夹 M10 六角螺母的加工	参照系	《1 型六角螺母》(GB/T 6170—2015)	
普适性工作过程 典型工作过程	资讯	计划	决策	实施	检查	评价
划端面正六边形	查看《1 型六角螺母》(GB/T 6170—2015),获取 M10 六角螺母的有关信息,便于安全有效地完成任务。	通过同学之间小组讨论,拟订用电安全规范计划,完成 3.1.4.2 划端面正六边形环节计划单。	计划的可行性、经济性、可操作性和实施难度几方面综合分析,选取最佳方案。	查找有关资料,完成 3.1.4.4 划端面正六边形环节实施单。	按照 3.1.4.5 划端面正六边形环节检查单的检查项目和标准,进行学生自查和教师检查。	根据评价项目配置分值,进行自评、互评和教师评价,完成 3.5 变压器低压套管线夹 M10 六角螺母的加工学习情境评价单。
锉正六棱面	查看《1 型六角螺母》(GB/T 6170—2015),获取 M10 六角螺母的有关信息,便于安全有效地完成任务。	通过同学之间小组讨论,拟订用电安全规范计划,完成 3.1.5.2 锉正六棱面环节计划单。	计划的可行性、经济性、可操作性和实施难度几方面综合分析,选取最佳方案。	查找有关资料,完成 3.1.5.4 锉正六棱面环节实施单。	按照 3.1.5.5 锉正六棱面环节检查单的检查项目和标准,进行学生自查和教师检查。	根据评价项目配置分值,进行自评、互评和教师评价,完成 3.5 变压器低压套管线夹 M10 六角螺母的加工学习情境评价单。
钻孔	查看《1 型六角螺母》(GB/T 6170—2015),获取 M10 六角螺母的有关信息,便于安全有效地完成任务。	通过同学之间小组讨论,拟订用电安全规范计划,完成 3.1.6.2 钻孔环节计划单。	计划的可行性、经济性、可操作性和实施难度几方面综合分析,选取最佳方案。	查找有关资料,完成 3.1.6.4 钻孔环节实施单。	按照 3.1.6.5 钻孔环节检查单的检查项目和标准,进行学生自查和教师检查。	根据评价项目配置分值,进行自评、互评和教师评价,完成 3.5 变压器低压套管线夹 M10 六角螺母的加工学习情境评价单。

续表

学习场	金工实训	学习情境 3	变压器低压套管线夹M10六角螺母的加工	参照系	《1 型六角螺母》（GB/T 6170—2015）	
普适性工作过程 典型工作过程	资讯	计划	决策	实施	检查	评价
攻螺纹	查看《1 型六角螺母》（GB/T 6170—2015），获取 M10 六角螺母的有关信息，便于安全有效地完成任务。	通过同学之间小组讨论，拟订用电安全规范计划，完成 3.1.7.2 攻螺纹环节计划单。	计划的可行性、经济性、可操作性和实施难度几方面综合分析，选取最佳方案。	查找有关资料，完成 3.1.7.4 攻螺纹环节实施单。	按照 3.1.7.5 攻螺纹环节检查单的检查项目和标准，进行学生自查和教师检查。	根据评价项目配置分值，进行自评、互评和教师评价，完成 3.5 变压器低压套管线夹 M10 六角螺母的加工学习情境评价单。
倒角	查看《1 型六角螺母》（GB/T 6170—2015），获取 M10 六角螺母的有关信息，便于安全有效地完成任务。	通过同学之间小组讨论，拟订用电安全规范计划，完成 3.1.8.2 倒角环节计划单。	计划的可行性、经济性、可操作性和实施难度几方面综合分析，选取最佳方案。	查找有关资料，完成 3.1.8.4 倒角环节实施单。	按照 3.1.8.5 倒角环节检查单的检查项目和标准，进行学生自查和教师检查。	根据评价项目配置分值，进行自评、互评和教师评价，完成 3.5 变压器低压套管线夹 M10 六角螺母的加工学习情境评价单。

学习情境 3-4

3.1.1　下料

3.1.1.1　下料环节资讯单

<table>
<tr><td>学习场</td><td colspan="3">金　工　实　训</td></tr>
<tr><td>学习情境 3</td><td>变压器低压套管线夹 M10 六角螺母的加工</td><td>本环节学时</td><td>6 分钟</td></tr>
<tr><td>典型工作过程描述</td><td colspan="3">下料—锯割—锉端面—划端面正六边形—锉正六棱面—钻孔—攻螺纹—倒角</td></tr>
<tr><td>搜集资讯的方式</td><td colspan="3">教师的讲解、互联网查询或者同学间交流</td></tr>
<tr><td>资讯描述</td><td colspan="3">查看教师提供的资料，获取 M10 六角螺母制作下料的有关信息，便于安全有效地完成任务，如图 3.1.1 所示。

图 3.1.1　线夹的 M10 六角螺母</td></tr>
<tr><td>对学生的要求</td><td colspan="3">课前充分预习实训作业单，充分利用网络查找有关实训的学习资料，实训过程中穿着实训服装，贯彻落实自己不伤害自己、自己不伤害他人、自己不被他人伤害、保护他人不被伤害的“四不伤害”安全要求，注意严格遵守实训室的各项规章制度。</td></tr>
<tr><td>参考资料</td><td colspan="3">参考《施工现场临时用电安全技术规范》（JGJ 46—2005）和《1 型六角螺母》（GB/T 6170—2015）。</td></tr>
</table>

学习情境 3-5

3.1.1.2 下料环节计划单

学习场	金 工 实 训		
学习情境 3	变压器低压套管线夹 M10 六角螺母的加工	本环节学时	6 分钟
典型工作过程描述	下料—锯割—锉端面—划端面正六边形—锉正六棱面—钻孔—攻螺纹—倒角		
计划制订的方式	同学之间小组讨论		

序号	工作步骤	注意事项
1	工作前准备	
2	开班前会，布置任务	
3	安全操作	
4	验收产品	

计划评价					
	班级		第（　）组	组长签字	
	教师签字		日期		
	评语：				

3.1.1.3　下料环节决策单

<table>
<tr><td>学习场</td><td colspan="5">金　工　实　训</td></tr>
<tr><td>学习情境 3</td><td colspan="3">变压器低压套管线夹 M10 六角螺母的加工</td><td>本环节学时</td><td>6 分钟</td></tr>
<tr><td>典型工作过程描述</td><td colspan="5">下料—锯割—锉端面—划端面正六边形—锉正六棱面—钻孔—攻螺纹—倒角</td></tr>
<tr><td colspan="6">计　划　对　比</td></tr>
<tr><td>序号</td><td>计划的可行性</td><td>计划的经济性</td><td>计划的可操作性</td><td>计划的实施难度</td><td>综合评价</td></tr>
<tr><td>方案 1</td><td></td><td></td><td></td><td></td><td></td></tr>
<tr><td>方案 2</td><td></td><td></td><td></td><td></td><td></td></tr>
<tr><td>方案 3</td><td></td><td></td><td></td><td></td><td></td></tr>
<tr><td rowspan="3">决策评价</td><td>班级</td><td></td><td>第（　　）组</td><td>组长签字</td><td></td></tr>
<tr><td>教师签字</td><td></td><td>日期</td><td colspan="2"></td></tr>
<tr><td colspan="5">评语：</td></tr>
</table>

学习情境 3-7

3.1.1.4 下料环节实施单

<table>
<tr><td>学习场</td><td colspan="4">金 工 实 训</td></tr>
<tr><td>学习情境 3</td><td colspan="2">变压器低压套管线夹 M10 六角螺母的加工</td><td>本环节学时</td><td>30 分钟</td></tr>
<tr><td>典型工作过程描述</td><td colspan="4">下料—锯割—锉端面—划端面正六边形—锉正六棱面—钻孔—攻螺纹—倒角</td></tr>
<tr><td>序号</td><td>实施步骤</td><td>工作内容及方法</td><td>工艺质量要求</td><td>注意事项</td></tr>
<tr><td>1</td><td>工作前准备</td><td></td><td></td><td></td></tr>
<tr><td>2</td><td>开班前会，布置任务</td><td></td><td></td><td></td></tr>
<tr><td>3</td><td>安全操作</td><td>利用直尺或者游标卡尺检查来料尺寸——ϕ25mm 圆钢。</td><td>ϕ25mm 圆钢的长度大于 10mm 且小于 15mm 端面饱满无缺陷。</td><td>铁屑伤害和锋利物伤害。</td></tr>
<tr><td>4</td><td>验收产品</td><td></td><td></td><td></td></tr>
<tr><td colspan="5">实施说明：</td></tr>
</table>

<table>
<tr><td rowspan="3">实施评价</td><td>班级</td><td></td><td>第（ ）组</td><td>组长签字</td><td></td></tr>
<tr><td>教师签字</td><td></td><td>日期</td><td colspan="2"></td></tr>
<tr><td colspan="5">评语：</td></tr>
</table>

学习情境 3-8

3.1.1.5　下料环节检查单

学习场	金　工　实　训				
学习情境 3	变压器低压套管线夹 M10 六角螺母的加工		本环节学时	6 分钟	
典型工作过程描述	下料—锯割—锉端面—划端面正六边形—锉正六棱面—钻孔—攻螺纹—倒角				
序号	检查项目	检 查 标 准	学生自查	教师检查	
1	下料资讯环节	获得 M10 六角螺母下料的有关信息的来源。			
2	下料计划环节	制定下料步骤的各个方案。			
3	下料决策环节	对各种下料的方案衡量对比评分，获取最佳方案。			
4	下料实施环节	下料过程中各个环节按照标准执行，贯彻落实“四不伤害”。			
5	下料检查环节	各个环节逐一检查，是否有爱岗敬业的思想。			
检查评价	班级		第（　）组	组长签字	
	教师签字		日期		
	评语：				

3.1.2　锯割

3.1.2.1　锯割环节资讯单

学习场	金　工　实　训		
学习情境 3	变压器低压套管线夹 M10 六角螺母的加工	本环节学时	6 分钟
典型工作过程描述	下料—锯割—锉端面—划端面正六边形—锉正六棱面—钻孔—攻螺纹—倒角		
搜集资讯的方式	教师的讲解、互联网查询或者同学间交流		
资讯描述	查看教师提供的资料，获取锯割环节的有关信息，便于安全有效地完成任务，如图 3.1.2 所示。  图 3.1.2　现场锯割		
对学生的要求	课前充分预习实训作业单，充分利用网络查找有关实训的学习资料，实训过程中穿着实训服装，贯彻落实自己不伤害自己、自己不伤害他人、自己不被他人伤害、保护他人不被伤害的“四不伤害”安全要求，注意严格遵守实训室的各项规章制度。		
参考资料	参考《1 型六角螺母》(GB/T 6170—2015)。 锯割的操作		

3.1.2.2　锯割环节计划单

<table>
<tr><td colspan="2">学习场</td><td colspan="5">金　工　实　训</td></tr>
<tr><td colspan="2">学习情境 3</td><td colspan="3">变压器低压套管线夹 M10 六角螺母的加工</td><td>本环节学时</td><td>6 分钟</td></tr>
<tr><td colspan="2">典型工作过程描述</td><td colspan="5">下料—锯割—锉端面—划端面正六边形—锉正六棱面—钻孔—攻螺纹—倒角</td></tr>
<tr><td colspan="2">计划制订的方式</td><td colspan="5">同学之间小组讨论</td></tr>
<tr><td>序号</td><td>工 作 步 骤</td><td colspan="5">安 全 文 明 生 产</td></tr>
<tr><td>1</td><td>检查锯弓好坏</td><td colspan="5"></td></tr>
<tr><td>2</td><td>装锯条</td><td colspan="5"></td></tr>
<tr><td>3</td><td>锯割</td><td colspan="5"></td></tr>
<tr><td>4</td><td>清理现场</td><td colspan="5"></td></tr>
<tr><td colspan="2" rowspan="3">计划评价</td><td>班级</td><td></td><td>第（　　）组</td><td>组长签字</td><td></td></tr>
<tr><td>教师签字</td><td></td><td>日期</td><td colspan="2"></td></tr>
<tr><td colspan="5">评语：</td></tr>
</table>

学习情境 3-11

不经历风雨，怎么见彩虹。苦练千百次，操作如有神。

3.1.2.3 锯割环节决策单

<table>
<tr><td>学习场</td><td colspan="6">金 工 实 训</td></tr>
<tr><td>学习情境 3</td><td colspan="3">变压器低压套管线夹 M10 六角螺母的加工</td><td colspan="2">本环节学时</td><td>6 分钟</td></tr>
<tr><td>典型工作过程描述</td><td colspan="6">下料—锯割—锉端面—划端面正六边形—锉正六棱面—钻孔—攻螺纹—倒角</td></tr>
<tr><td colspan="7">计 划 对 比</td></tr>
<tr><td>序号</td><td>计划的可行性</td><td>计划的经济性</td><td colspan="2">计划的可操作性</td><td>计划的实施难度</td><td>综合评价</td></tr>
<tr><td>方案 1</td><td></td><td></td><td colspan="2"></td><td></td><td></td></tr>
<tr><td>方案 2</td><td></td><td></td><td colspan="2"></td><td></td><td></td></tr>
<tr><td>方案 3</td><td></td><td></td><td colspan="2"></td><td></td><td></td></tr>
<tr><td rowspan="3">决策评价</td><td>班级</td><td></td><td>第（ ）组</td><td>组长签字</td><td colspan="2"></td></tr>
<tr><td>教师签字</td><td></td><td>日期</td><td colspan="3"></td></tr>
<tr><td colspan="6">评语：</td></tr>
</table>

学习情境 3-12

3.1.2.4　锯割环节实施单

<table>
<tr><td colspan="2">学习场</td><td colspan="3">金　工　实　训</td></tr>
<tr><td colspan="2">学习情境 3</td><td colspan="2">变压器低压套管线夹 M10 六角螺母的加工</td><td>本环节学时 / 30 分钟</td></tr>
<tr><td colspan="2">典型工作过程描述</td><td colspan="3">下料—锯割—锉端面—划端面正六边形—锉正六棱面—钻孔—攻螺纹—倒角</td></tr>
<tr><td>序号</td><td>实施步骤</td><td>工作内容及方法</td><td>工艺质量要求</td><td>安全文明生产</td></tr>
<tr><td>1</td><td>检查锯弓好坏</td><td></td><td></td><td></td></tr>
<tr><td>2</td><td>装锯条</td><td></td><td></td><td></td></tr>
<tr><td>3</td><td>安全操作锯割</td><td>以锉平的端面 A［图 3.0.1（a）］为基准面锯下 10mm 长的钢料。</td><td>锯下端面 A［图 3.0.1（a）］的对面要基本平整，长度为 10mm 的钢料。</td><td>锯齿伤害。</td></tr>
<tr><td>4</td><td>清理现场</td><td></td><td></td><td></td></tr>
<tr><td colspan="5">实施说明：</td></tr>
<tr><td rowspan="3">实施评价</td><td>班级</td><td></td><td>第（　　）组</td><td>组长签字</td></tr>
<tr><td>教师签字</td><td></td><td>日期</td><td></td></tr>
<tr><td colspan="4">评语：</td></tr>
</table>

学习情境 3-13

3.1.2.5　锯割环节检查单

<table>
<tr><td>学习场</td><td colspan="6">金　工　实　训</td></tr>
<tr><td>学习情境 3</td><td colspan="3">变压器低压套管线夹 M10 六角螺母的加工</td><td colspan="2">本环节学时</td><td>6 分钟</td></tr>
<tr><td>典型工作过程描述</td><td colspan="6">下料—锯割—锉端面—划端面正六边形—锉正六棱面—钻孔—攻螺纹—倒角</td></tr>
<tr><td>序号</td><td>检 查 项 目</td><td colspan="2">检 查 标 准</td><td colspan="2">学生自查</td><td>教师检查</td></tr>
<tr><td>1</td><td>锯割资讯环节</td><td colspan="2">获得 M10 六角螺母锯割的有关信息的来源。</td><td colspan="2"></td><td></td></tr>
<tr><td>2</td><td>锯割计划环节</td><td colspan="2">制定锯割步骤的各个方案。</td><td colspan="2"></td><td></td></tr>
<tr><td>3</td><td>锯割决策环节</td><td colspan="2">对各种锯割的方案衡量对比评分，获取最佳方案。</td><td colspan="2"></td><td></td></tr>
<tr><td>4</td><td>锯割实施环节</td><td colspan="2">锯割过程中各个环节按照标准执行，贯彻落实“四不伤害”。</td><td colspan="2"></td><td></td></tr>
<tr><td>5</td><td>锯割检查环节</td><td colspan="2">各个环节逐一检查，是否有爱岗敬业的主题思想。</td><td colspan="2"></td><td></td></tr>
<tr><td rowspan="3">检查评价</td><td>班级</td><td></td><td>第（　　）组</td><td>组长签字</td><td colspan="2"></td></tr>
<tr><td>教师签字</td><td></td><td>日期</td><td colspan="3"></td></tr>
<tr><td colspan="6">评语：</td></tr>
</table>

3.1.3　锉端面

3.1.3.1　锉端面环节资讯单

学习场	金　工　实　训		
学习情境 3	变压器低压套管线夹 M10 六角螺母的加工	本环节学时	6 分钟
典型工作过程描述	下料—锯割—锉端面—划端面正六边形—锉正六棱面—钻孔—攻螺纹—倒角		
搜集资讯的方式	教师的讲解、互联网查询或者同学间交流		
资讯描述	查看教师提供的资料，获取锉削的有关信息，便于安全有效地完成任务，如图 3.1.3 所示。  图 3.1.3　现场锉削		
对学生的要求	课前充分预习实训作业单，充分利用网络查找有关实训的学习资料，实训过程中穿着实训服装，贯彻落实自己不伤害自己、自己不伤害他人、自己不被他人伤害、保护他人不被伤害的“四不伤害”安全要求，注意严格遵守实训室的各项规章制度。		
参考资料	参考《1 型六角螺母》(GB/T 6170—2015)。 锉削的操作		

3.1.3.2 锉端面环节计划单

<table>
<tr><td>学习场</td><td colspan="5">金 工 实 训</td></tr>
<tr><td>学习情境 3</td><td colspan="3">变压器低压套管线夹 M10 六角螺母的加工</td><td>本环节学时</td><td>6 分钟</td></tr>
<tr><td>典型工作过程描述</td><td colspan="5">下料—锯割—锉端面—划端面正六边形—锉正六棱面—钻孔—攻螺纹—倒角</td></tr>
<tr><td>计划制订的方式</td><td colspan="5">同学之间小组讨论</td></tr>
<tr><td>序号</td><td colspan="2">工作步骤</td><td colspan="3">注意事项</td></tr>
<tr><td>1</td><td colspan="2">工作前准备</td><td colspan="3"></td></tr>
<tr><td>2</td><td colspan="2">开班前会，布置任务</td><td colspan="3"></td></tr>
<tr><td>3</td><td colspan="2">安全操作</td><td colspan="3">铁屑伤害和锋利物伤害。</td></tr>
<tr><td>4</td><td colspan="2">漏光法检查平面是否平整</td><td colspan="3"></td></tr>
<tr><td>5</td><td colspan="2">清理现场</td><td colspan="3"></td></tr>
<tr><td rowspan="3">计划评价</td><td>班级</td><td></td><td>第（　　）组</td><td>组长签字</td><td></td></tr>
<tr><td>教师签字</td><td></td><td>日期</td><td colspan="2"></td></tr>
<tr><td colspan="5">评语：</td></tr>
</table>

学习情境 3-16

3.1.3.3　锉端面环节决策单

<table>
<tr><td colspan="2">学习场</td><td colspan="4">金　工　实　训</td></tr>
<tr><td colspan="2">学习情境 3</td><td colspan="2">变压器低压套管线夹 M10 六角螺母的加工</td><td>本环节学时</td><td>6 分钟</td></tr>
<tr><td colspan="2">典型工作过程描述</td><td colspan="4">下料—锯割—锉端面—划端面正六边形—锉正六棱面—钻孔—攻螺纹—倒角</td></tr>
<tr><td colspan="6">计　划　对　比</td></tr>
<tr><td>序号</td><td>计划的可行性</td><td>计划的经济性</td><td>计划的可操作性</td><td>计划的实施难度</td><td>综合评价</td></tr>
<tr><td>方案 1</td><td></td><td></td><td></td><td></td><td></td></tr>
<tr><td>方案 2</td><td></td><td></td><td></td><td></td><td></td></tr>
<tr><td>方案 3</td><td></td><td></td><td></td><td></td><td></td></tr>
<tr><td rowspan="3">决策评价</td><td>班级</td><td></td><td>第（　　）组</td><td>组长签字</td><td></td></tr>
<tr><td>教师签字</td><td></td><td>日期</td><td colspan="2"></td></tr>
<tr><td colspan="5">评语：</td></tr>
</table>

3.1.3.4 锉端面环节实施单

<table>
<tr><td colspan="2">学习场</td><td colspan="3">金 工 实 训</td></tr>
<tr><td colspan="2">学习情境 3</td><td>变压器低压套管线夹 M10 六角螺母的加工</td><td>本环节学时</td><td>60 分钟</td></tr>
<tr><td colspan="2">典型工作过程描述</td><td colspan="3">下料—锯割—锉端面—划端面正六边形—锉正六棱面—钻孔—攻螺纹—倒角</td></tr>
<tr><td>序号</td><td>实施步骤</td><td>工作内容及方法</td><td>工艺质量要求</td><td>注意事项</td></tr>
<tr><td>1</td><td>工作前准备</td><td></td><td></td><td></td></tr>
<tr><td>2</td><td>开班前会，布置任务</td><td></td><td></td><td></td></tr>
<tr><td>3</td><td>安全操作</td><td>以端面 A［图 3.0.1（a）］为基准，粗锉、精锉相对面，达到尺寸公差、平行度和表面粗糙度的要求。</td><td>利用角尺进行漏光法检查端面 B［图 3.0.1（a）］粗糙度，要求两端面高度为 8.4mm。</td><td>锋利物伤害和铁屑伤害。</td></tr>
<tr><td>4</td><td>漏光法检查平面是否平整</td><td></td><td></td><td></td></tr>
<tr><td>5</td><td>清理现场</td><td></td><td></td><td></td></tr>
<tr><td colspan="5">实施说明：</td></tr>
<tr><td colspan="2" rowspan="3">实施评价</td><td>班级</td><td>第（　　）组</td><td>组长签字</td></tr>
<tr><td>教师签字</td><td colspan="2">日期</td></tr>
<tr><td colspan="3">评语：</td></tr>
</table>

3.1.3.5　锉端面环节检查单

<table>
<tr><td colspan="2">学习场</td><td colspan="5">金　工　实　训</td></tr>
<tr><td colspan="2">学习情境 3</td><td colspan="2">变压器低压套管线夹 M10 六角螺母的加工</td><td colspan="2">本环节学时</td><td>6 分钟</td></tr>
<tr><td colspan="2">典型工作过程描述</td><td colspan="5">下料—锯割—锉端面—划端面正六边形—锉正六棱面—钻孔—攻螺纹—倒角</td></tr>
<tr><td>序号</td><td>检查项目</td><td colspan="2">检 查 标 准</td><td colspan="2">学生自查</td><td>教师检查</td></tr>
<tr><td>1</td><td>锉端面资讯环节</td><td colspan="2">获得 M10 六角螺母锉削的有关信息的来源。</td><td colspan="2"></td><td></td></tr>
<tr><td>2</td><td>锉端面计划环节</td><td colspan="2">制定锉削步骤的各个方案。</td><td colspan="2"></td><td></td></tr>
<tr><td>3</td><td>锉端面决策环节</td><td colspan="2">对各种锉削的方案衡量对比评分，获取最佳方案。</td><td colspan="2"></td><td></td></tr>
<tr><td>4</td><td>锉端面实施环节</td><td colspan="2">锉削过程中各个环节按照标准执行，贯彻落实“四不伤害”。</td><td colspan="2"></td><td></td></tr>
<tr><td>5</td><td>锉端面检查环节</td><td colspan="2">各个环节逐一检查，是否有爱岗敬业的思想。</td><td colspan="2"></td><td></td></tr>
<tr><td colspan="2" rowspan="3">检查评价</td><td>班级</td><td></td><td>第（　　）组</td><td>组长签字</td><td></td></tr>
<tr><td>教师签字</td><td></td><td>日期</td><td colspan="2"></td></tr>
<tr><td colspan="5">评语：</td></tr>
</table>

3.1.4 划端面正六边形

3.1.4.1 划端面正六边形环节资讯单

学习场	金工实训		
学习情境 3	变压器低压套管线夹 M10 六角螺母的加工	本环节学时	6 分钟
典型工作过程描述	下料—锯割—锉端面—划端面正六边形—锉正六棱面—钻孔—攻螺纹—倒角		
搜集资讯的方式	教师的讲解、互联网查询或者同学间交流		
资讯描述	查看教师提供的资料，获取六角螺母制作划线的有关信息，便于安全有效地完成任务，参考图 3.1.4 划线。 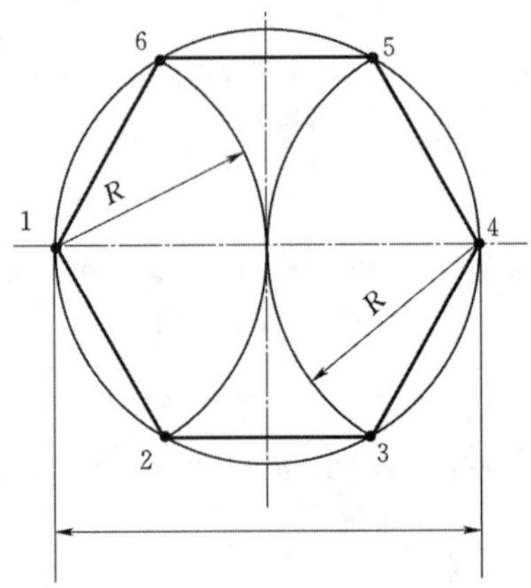 图 3.1.4 划端面正六边形		
对学生的要求	课前充分预习实训作业单，充分利用网络查找有关实训的学习资料，实训过程中穿着实训服装，贯彻落实自己不伤害自己、自己不伤害他人、自己不被他人伤害、保护他人不被伤害的“四不伤害”安全要求，注意严格遵守实训室的各项规章制度。		
参考资料	参考《1 型六角螺母》(GB/T 6170—2015)。		

3.1.4.2　划端面正六边形环节计划单

<table>
<tr><td>学习场</td><td colspan="6">金　工　实　训</td></tr>
<tr><td>学习情境 3</td><td colspan="4">变压器低压套管线夹 M10 六角螺母的加工</td><td>本环节学时</td><td>6 分钟</td></tr>
<tr><td>典型工作过程描述</td><td colspan="6">下料—锯割—锉端面—划端面正六边形—锉正六棱面—钻孔—攻螺纹—倒角</td></tr>
<tr><td>计划制订的方式</td><td colspan="6">同学之间小组讨论</td></tr>
<tr><td>序号</td><td colspan="2">工作步骤</td><td colspan="4">注意事项</td></tr>
<tr><td>1</td><td colspan="2">工作前准备</td><td colspan="4"></td></tr>
<tr><td>2</td><td colspan="2">开班前会，布置任务</td><td colspan="4"></td></tr>
<tr><td>3</td><td colspan="2">安全操作划线</td><td colspan="4"></td></tr>
<tr><td>4</td><td colspan="2">整理现场</td><td colspan="4"></td></tr>
<tr><td rowspan="3">计划评价</td><td>班级</td><td></td><td>第（　　）组</td><td>组长签字</td><td colspan="2"></td></tr>
<tr><td>教师签字</td><td></td><td>日期</td><td colspan="3"></td></tr>
<tr><td colspan="6">评语：</td></tr>
</table>

学习情境 3-21

3.1.4.3 划端面正六边形环节决策单

<table>
<tr><td>学习场</td><td colspan="6">金 工 实 训</td></tr>
<tr><td>学习情境 3</td><td colspan="3">变压器低压套管线夹 M10 六角螺母的加工</td><td colspan="2">本环节学时</td><td>6 分钟</td></tr>
<tr><td>典型工作过程描述</td><td colspan="6">下料—锯割—锉端面—划端面正六边形—锉正六棱面—钻孔—攻螺纹—倒角</td></tr>
<tr><td colspan="7">计 划 对 比</td></tr>
<tr><td>序号</td><td>计划的可行性</td><td>计划的经济性</td><td>计划的可操作性</td><td colspan="2">计划的实施难度</td><td>综合评价</td></tr>
<tr><td>方案 1</td><td></td><td></td><td></td><td colspan="2"></td><td></td></tr>
<tr><td>方案 2</td><td></td><td></td><td></td><td colspan="2"></td><td></td></tr>
<tr><td>方案 3</td><td></td><td></td><td></td><td colspan="2"></td><td></td></tr>
<tr><td rowspan="3">决策评价</td><td>班级</td><td></td><td>第（ ）组</td><td>组长签字</td><td colspan="2"></td></tr>
<tr><td>教师签字</td><td></td><td>日期</td><td colspan="3"></td></tr>
<tr><td colspan="6">评语：</td></tr>
</table>

3.1.4.4　划端面正六边形环节实施单

<table>
<tr><td>学习场</td><td colspan="4">金　工　实　训</td></tr>
<tr><td>学习情境 3</td><td colspan="2">变压器低压套管线夹 M10 六角螺母的加工</td><td>本环节学时</td><td>30 分钟</td></tr>
<tr><td>典型工作过程描述</td><td colspan="4">下料—锯割—锉端面—划端面正六边形—锉正六棱面—钻孔—攻螺纹—倒角</td></tr>
</table>

序号	实施步骤	工作内容及方法	工艺质量要求	注意事项
1	工作前准备			
2	开班前会，布置任务			
3	安全操作划线	利用圆规、划针、样冲和卡尺划六角螺母的正六边形的加工界线。	要求正六边形对边距离为 16mm。	锋利物伤害。
4	清理现场			

实施说明：

<table>
<tr><td rowspan="3">实施评价</td><td>班级</td><td></td><td>第（　　）组</td><td>组长签字</td><td></td></tr>
<tr><td>教师签字</td><td></td><td>日期</td><td colspan="2"></td></tr>
<tr><td colspan="5">评语：</td></tr>
</table>

3.1.4.5 划端面正六边形环节检查单

<table>
<tr><td colspan="2">学习场</td><td colspan="4">金 工 实 训</td></tr>
<tr><td colspan="2">学习情境 3</td><td colspan="2">变压器低压套管线夹 M10 六角螺母的加工</td><td>本环节学时</td><td>6 分钟</td></tr>
<tr><td colspan="2">典型工作过程描述</td><td colspan="4">下料—锯割—锉端面—划端面正六边形—锉正六棱面—钻孔—攻螺纹—倒角</td></tr>
<tr><td>序号</td><td>检查项目</td><td colspan="2">检 查 标 准</td><td>学生自查</td><td>教师检查</td></tr>
<tr><td>1</td><td>资讯环节</td><td colspan="2">获得 M10 六角螺母制作中划线的有关信息的来源。</td><td></td><td></td></tr>
<tr><td>2</td><td>计划环节</td><td colspan="2">制定划线步骤的各个方案。</td><td></td><td></td></tr>
<tr><td>3</td><td>决策环节</td><td colspan="2">对各种划线的方案衡量对比评分，获取最佳方案。</td><td></td><td></td></tr>
<tr><td>4</td><td>实施环节</td><td colspan="2">划线过程中各个环节按照标准执行，贯彻落实“四不伤害”。</td><td></td><td></td></tr>
<tr><td>5</td><td>检查环节</td><td colspan="2">各个环节逐一检查，是否有爱岗敬业的思想。</td><td></td><td></td></tr>
<tr><td colspan="2" rowspan="3">检查评价</td><td>班级</td><td></td><td>第（ ）组</td><td>组长签字</td></tr>
<tr><td>教师签字</td><td></td><td>日期</td><td></td></tr>
<tr><td colspan="4">评语：</td></tr>
</table>

学习情境 3-24

不经历风雨，怎么见彩虹。苦练千百次，操作如有神。

3.1.5　锉正六棱面

3.1.5.1　锉正六棱面环节资讯单

<table>
<tr><td>学习场</td><td colspan="3">金　工　实　训</td></tr>
<tr><td>学习情境 3</td><td>变压器低压套管线夹 M10 六角螺母的加工</td><td>本环节学时</td><td>8 分钟</td></tr>
<tr><td>典型工作过程描述</td><td colspan="3">下料—锯割—锉端面—划端面正六边形—锉正六棱面—钻孔—攻螺纹—倒角</td></tr>
<tr><td>搜集资讯的方式</td><td colspan="3">教师的讲解、互联网查询或者同学间交流</td></tr>
<tr><td>资讯描述</td><td colspan="3">查看教师提供的资料，获取正六棱面制作的有关信息，便于安全有效地完成任务，参考图 3.1.5。

图 3.1.5　锉正六棱面</td></tr>
<tr><td>对学生的要求</td><td colspan="3">课前充分预习实训作业单，充分利用网络查找有关实训的学习资料，实训过程中穿着实训服装，贯彻落实自己不伤害自己、自己不伤害他人、自己不被他人伤害、保护他人不被伤害的“四不伤害”安全要求，注意严格遵守实训室的各项规章制度。</td></tr>
<tr><td>参考资料</td><td colspan="3">参考《1 型六角螺母》(GB/T 6170—2015)。</td></tr>
</table>

不经历风雨，怎么见彩虹。苦练千百次，操作如有神。

3.1.5.2 锉正六棱面环节计划单

<table>
<tr><td>学习场</td><td colspan="5">金 工 实 训</td></tr>
<tr><td>学习情境 3</td><td colspan="3">变压器低压套管线夹 M10 六角螺母的加工</td><td>本环节学时</td><td>8 分钟</td></tr>
<tr><td>典型工作过程描述</td><td colspan="5">下料—锯割—锉端面—划端面正六边形—锉正六棱面—钻孔—攻螺纹—倒角</td></tr>
<tr><td>计划制订的方式</td><td colspan="5">同学之间小组讨论</td></tr>
<tr><td>序号</td><td colspan="2">工 作 步 骤</td><td colspan="3">注 意 事 项</td></tr>
<tr><td>1</td><td colspan="2">工作前准备</td><td colspan="3"></td></tr>
<tr><td>2</td><td colspan="2">开班前会，布置任务</td><td colspan="3"></td></tr>
<tr><td>3</td><td colspan="2">安全操作锉正六棱面</td><td colspan="3"></td></tr>
<tr><td>4</td><td colspan="2">清理现场</td><td colspan="3"></td></tr>
<tr><td rowspan="3">计划评价</td><td>班级</td><td></td><td>第（ ）组</td><td>组长签字</td><td></td></tr>
<tr><td>教师签字</td><td></td><td>日期</td><td colspan="2"></td></tr>
<tr><td colspan="5">评语：</td></tr>
</table>

学习情境 3-26

3.1.5.3　锉正六棱面环节决策单

<table>
<tr><td>学习场</td><td colspan="6">金　工　实　训</td></tr>
<tr><td>学习情境 3</td><td colspan="3">变压器低压套管线夹 M10 六角螺母的加工</td><td colspan="2">本环节学时</td><td>8 分钟</td></tr>
<tr><td>典型工作过程描述</td><td colspan="6">下料—锯割—锉端面—划端面正六边形—锉正六棱面—钻孔—攻螺纹—倒角</td></tr>
<tr><td colspan="7">计　划　对　比</td></tr>
<tr><td>序号</td><td>计划的可行性</td><td>计划的经济性</td><td>计划的可操作性</td><td colspan="2">计划的实施难度</td><td>综合评价</td></tr>
<tr><td>方案 1</td><td></td><td></td><td></td><td colspan="2"></td><td></td></tr>
<tr><td>方案 2</td><td></td><td></td><td></td><td colspan="2"></td><td></td></tr>
<tr><td>方案 3</td><td></td><td></td><td></td><td colspan="2"></td><td></td></tr>
<tr><td rowspan="3">决策评价</td><td>班级</td><td></td><td>第（　　）组</td><td>组长签字</td><td colspan="2"></td></tr>
<tr><td>教师签字</td><td></td><td>日期</td><td colspan="3"></td></tr>
<tr><td colspan="6">评语：</td></tr>
</table>

学习情境 3-27

3.1.5.4 锉正六棱面环节实施单

<table>
<tr><td colspan="2">学习场</td><td colspan="4">金 工 实 训</td></tr>
<tr><td colspan="2">学习情境 3</td><td colspan="2">变压器低压套管线夹 M10 六角螺母的加工</td><td>本环节学时</td><td>100 分钟</td></tr>
<tr><td colspan="2">典型工作过程描述</td><td colspan="4">下料—锯割—锉端面—划端面正六边形—锉正六棱面—钻孔—攻螺纹—倒角</td></tr>
<tr><td>序号</td><td>实施步骤</td><td>工作内容及方法</td><td colspan="2">工艺质量要求</td><td>注意事项</td></tr>
<tr><td>1</td><td>工作前准备</td><td></td><td colspan="2"></td><td></td></tr>
<tr><td>2</td><td>开班前会，布置任务</td><td></td><td colspan="2"></td><td></td></tr>
<tr><td>3</td><td>安全操作锉正六棱面</td><td>沿六角螺母端面的正六边形的加工界线进行锉削，并沿界线锉成正六边形。</td><td colspan="2">要求正六边形对边距离为 16mm。</td><td>锋利物伤害。</td></tr>
<tr><td>4</td><td>清理现场</td><td></td><td colspan="2"></td><td></td></tr>
<tr><td>5</td><td></td><td></td><td colspan="2"></td><td></td></tr>
<tr><td colspan="6">实施说明：</td></tr>
</table>

<table>
<tr><td rowspan="3">实施评价</td><td>班级</td><td></td><td>第（　　）组</td><td>组长签字</td><td></td></tr>
<tr><td>教师签字</td><td></td><td>日期</td><td colspan="2"></td></tr>
<tr><td colspan="5">评语：</td></tr>
</table>

学习情境 3-28

3.1.5.5　锉正六棱面环节检查单

<table>
<tr><td colspan="2">学习场</td><td colspan="5">金　工　实　训</td></tr>
<tr><td colspan="2">学习情境 3</td><td colspan="3">变压器低压套管线夹 M10 六角螺母的加工</td><td>本环节学时</td><td>8 分钟</td></tr>
<tr><td colspan="2">典型工作过程描述</td><td colspan="5">下料—锯割—锉端面—划端面正六边形—锉正六棱面—钻孔—攻螺纹—倒角</td></tr>
<tr><td>序号</td><td>检查项目</td><td colspan="3">检 查 标 准</td><td>学生自查</td><td>教师检查</td></tr>
<tr><td>1</td><td>资讯环节</td><td colspan="3">获得正六棱面制作的有关信息的来源。</td><td></td><td></td></tr>
<tr><td>2</td><td>计划环节</td><td colspan="3">制定正六棱面制作步骤的各个方案。</td><td></td><td></td></tr>
<tr><td>3</td><td>决策环节</td><td colspan="3">对各种正六棱面制作的方案衡量对比评分，获取最佳方案。</td><td></td><td></td></tr>
<tr><td>4</td><td>实施环节</td><td colspan="3">正六棱面制作过程中各个环节按照标准执行，贯彻落实“四不伤害”。</td><td></td><td></td></tr>
<tr><td>5</td><td>检查环节</td><td colspan="3">各个环节逐一检查，是否有爱岗敬业的主题思想。</td><td></td><td></td></tr>
<tr><td colspan="2" rowspan="3">检查评价</td><td>班级</td><td></td><td>第（　　）组</td><td>组长签字</td><td></td></tr>
<tr><td>教师签字</td><td></td><td>日期</td><td colspan="2"></td></tr>
<tr><td colspan="5">评语：</td></tr>
</table>

3.1.6 钻孔

3.1.6.1 钻孔环节资讯单

学习场	金工实训		
学习情境3	变压器低压套管线夹M10六角螺母的加工	本环节学时	6分钟
典型工作过程描述	下料—锯割—锉端面—划端面正六边形—锉正六棱面—钻孔—攻螺纹—倒角		
搜集资讯的方式	教师的讲解、互联网查询或者同学间交流		
资讯描述	查看教师提供的资料，获取安全操作钻孔的有关信息，便于安全有效地完成任务，参考图3.1.6。		
对学生的要求	课前充分预习实训作业单，充分利用网络查找有关实训的学习资料，实训过程中穿着实训服装，贯彻落实自己不伤害自己、自己不伤害他人、自己不被他人伤害、保护他人不被伤害的“四不伤害”安全要求，注意严格遵守实训室的各项规章制度。		
参考资料	参考《1型六角螺母》(GB/T 6170—2015)。		

图3.1.6为钻削施工现场，请指出违规操作的地方：____

图3.1.6 现场操作钻削

3.1.6.2　钻孔环节计划单

<table>
<tr><td>学习场</td><td colspan="5">金　工　实　训</td></tr>
<tr><td>学习情境 3</td><td colspan="3">变压器低压套管线夹 M10 六角螺母的加工</td><td>本环节学时</td><td>6 分钟</td></tr>
<tr><td>典型工作过程描述</td><td colspan="5">下料—锯割—锉端面—划端面正六边形—锉正六棱面—钻孔—攻螺纹—倒角</td></tr>
<tr><td>计划制订的方式</td><td colspan="5">同学之间小组讨论</td></tr>
<tr><td>序号</td><td colspan="3">工 作 步 骤</td><td colspan="2">注 意 事 项</td></tr>
<tr><td>1</td><td colspan="3">工作前准备</td><td colspan="2"></td></tr>
<tr><td>2</td><td colspan="3">开班前会，布置任务</td><td colspan="2"></td></tr>
<tr><td>3</td><td colspan="3">安全操作钻孔</td><td colspan="2">触电和钻头伤害。</td></tr>
<tr><td>4</td><td colspan="3">清理现场</td><td colspan="2"></td></tr>
<tr><td rowspan="3">计划评价</td><td>班级</td><td></td><td>第（　　）组</td><td>组长签字</td><td></td></tr>
<tr><td>教师签字</td><td></td><td>日期</td><td colspan="2"></td></tr>
<tr><td colspan="5">评语：</td></tr>
</table>

不经历风雨，怎么见彩虹。苦练千百次，操作如有神。

3.1.6.3　钻孔环节决策单

<table>
<tr><td colspan="2">学习场</td><td colspan="4">金　工　实　训</td></tr>
<tr><td colspan="2">学习情境 3</td><td colspan="2">变压器低压套管线夹 M10 六角螺母的加工</td><td>本环节学时</td><td>6 分钟</td></tr>
<tr><td colspan="2">典型工作过程描述</td><td colspan="4">下料—锯割—锉端面—划端面正六边形—锉正六棱面—钻孔—攻螺纹—倒角</td></tr>
<tr><td colspan="6">计　划　对　比</td></tr>
<tr><td>序号</td><td>计划的可行性</td><td>计划的经济性</td><td>计划的可操作性</td><td>计划的实施难度</td><td>综合评价</td></tr>
<tr><td>方案 1</td><td></td><td></td><td></td><td></td><td></td></tr>
<tr><td>方案 2</td><td></td><td></td><td></td><td></td><td></td></tr>
<tr><td>方案 3</td><td></td><td></td><td></td><td></td><td></td></tr>
<tr><td rowspan="3">决策评价</td><td>班级</td><td></td><td>第（　　）组</td><td>组长签字</td><td></td></tr>
<tr><td>教师签字</td><td></td><td>日期</td><td colspan="2"></td></tr>
<tr><td colspan="5">评语：</td></tr>
</table>

学习情境 3-32

3.1.6.4 钻孔环节实施单

<table>
<tr><td colspan="2">学习场</td><td colspan="3">金　工　实　训</td></tr>
<tr><td colspan="2">学习情境 3</td><td colspan="2">变压器低压套管线夹 M10 六角螺母的加工</td><td>本环节学时 20 分钟</td></tr>
<tr><td colspan="2">典型工作过程描述</td><td colspan="3">下料—锯割—锉端面—划端面正六边形—锉正六棱面—钻孔—攻螺纹—倒角</td></tr>
<tr><td>序号</td><td>实施步骤</td><td>工作内容及方法</td><td>工艺质量要求</td><td>注意事项</td></tr>
<tr><td>1</td><td>工作前准备</td><td></td><td></td><td></td></tr>
<tr><td>2</td><td>开班前会，布置任务</td><td></td><td></td><td></td></tr>
<tr><td>3</td><td>安全操作钻孔</td><td>在圆心打好样冲眼，用 ϕ8.5 钻头钻孔。用钻头 ϕ12.5 扩孔。</td><td>样冲眼要求深度 1.5mm、钻孔在圆心位置。两端面的孔口位置各扩孔深度 1.5mm。</td><td>触电和钻头伤害。</td></tr>
<tr><td>4</td><td>清理现场</td><td></td><td></td><td></td></tr>
<tr><td colspan="5">实施说明：</td></tr>
</table>

<table>
<tr><td rowspan="3">实施评价</td><td>班级</td><td></td><td>第（　　）组</td><td>组长签字</td><td></td></tr>
<tr><td>教师签字</td><td></td><td>日期</td><td colspan="2"></td></tr>
<tr><td colspan="5">评语：</td></tr>
</table>

学习情境 3-33

3.1.6.5 钻孔环节检查单

学习场	金工实训		
学习情境 3	变压器低压套管线夹 M10 六角螺母的加工	本环节学时	6 分钟
典型工作过程描述	下料—锯割—锉端面—划端面正六边形—锉正六棱面—钻孔—攻螺纹—倒角		

序号	检查项目	检查标准	学生自查	教师检查
1	钻孔资讯环节	获得 M10 六角螺母钻孔的有关信息的来源。		
2	钻孔计划环节	制定钻孔步骤的各个方案。		
3	钻孔决策环节	对各种钻孔的方案衡量对比评分，获取最佳方案。		
4	钻孔实施环节	钻孔过程中各个环节按照标准执行，贯彻落实“四不伤害”。		
5	钻孔检查环节	各个环节逐一检查，是否有爱岗敬业的主题思想。		

检查评价	班级		第（ ）组	组长签字	
	教师签字		日期		
	评语：				

3.1.7　攻螺纹

3.1.7.1　攻螺纹环节资讯单

学习场	金　工　实　训		
学习情境 3	变压器低压套管线夹 M10 六角螺母的加工	本环节学时	6 分钟
典型工作过程描述	下料—锯割—锉端面—划端面正六边形—锉正六棱面—钻孔—攻螺纹—倒角		
搜集资讯的方式	教师的讲解、互联网查询或者同学间交流		
资讯描述	查看教师提供的资料，获取六角螺母攻螺纹制作的有关信息，便于安全有效地完成任务。		
对学生的要求	课前充分预习实训作业单，充分利用网络查找有关实训的学习资料，实训过程中穿着实训服装，贯彻落实自己不伤害自己、自己不伤害他人、自己不被他人伤害、保护他人不被伤害的“四不伤害”安全要求，注意严格遵守实训室的各项规章制度。		
参考资料	参考《1 型六角螺母》(GB/T 6170—2015) 和《施工现场临时用电安全技术规范》(JGJ 46—2005)。 攻螺纹的操作		

技　能　要　点

下列说法正确的是（　　　　　　　　）。

A. 攻螺纹时保持丝锥垂直于螺母的端面，丝锥中心与孔中心重合。

B. 双手握紧铰杠两端，顺时针转动铰杠，每转动 4～5 圈要反转 1～2 圈。

C. 攻螺纹时先用一锥，然后再用二锥。

D. 丝锥与孔面不垂直时可以用小锤敲打丝锥。

E. 攻螺纹感到阻力很大时，可以强行用力扭动丝锥使其继续前进。

F. 锉削的速度一般为每分钟 30～60 次。

G. 锯割的速度一般为每分钟 60 次。

H. 钻孔时不能戴手套操作。

I. 更换钻头时必须断开台钻的电源。

3.1.7.2 攻螺纹环节计划单

<table>
<tr><td>学习场</td><td colspan="5">金 工 实 训</td></tr>
<tr><td>学习情境 3</td><td colspan="3">变压器低压套管线夹 M10 六角螺母的加工</td><td>本环节学时</td><td>6 分钟</td></tr>
<tr><td>典型工作过程描述</td><td colspan="5">下料—锯割—锉端面—划端面正六边形—锉正六棱面—钻孔—攻螺纹—倒角</td></tr>
<tr><td>计划制订的方式</td><td colspan="5">同学之间小组讨论</td></tr>
<tr><td>序号</td><td colspan="2">工 作 步 骤</td><td colspan="3">注 意 事 项</td></tr>
<tr><td>1</td><td colspan="2">工作前准备</td><td colspan="3"></td></tr>
<tr><td>2</td><td colspan="2">开班前会，布置任务</td><td colspan="3"></td></tr>
<tr><td>3</td><td colspan="2">安全操作攻螺纹</td><td colspan="3"></td></tr>
<tr><td>4</td><td colspan="2">清理现场</td><td colspan="3"></td></tr>
<tr><td rowspan="3">计划评价</td><td>班级</td><td></td><td>第（ ）组</td><td>组长签字</td><td></td></tr>
<tr><td>教师签字</td><td></td><td>日期</td><td colspan="2"></td></tr>
<tr><td colspan="5">评语：</td></tr>
</table>

学习情境 3-36

3.1.7.3　攻螺纹环节决策单

<table>
<tr><td>学习场</td><td colspan="5">金　工　实　训</td></tr>
<tr><td>学习情境 3</td><td colspan="3">变压器低压套管线夹 M10 六角螺母的加工</td><td>本环节学时</td><td>6 分钟</td></tr>
<tr><td>典型工作过程描述</td><td colspan="5">下料—锯割—锉端面—划端面正六边形—锉正六棱面—钻孔—攻螺纹—倒角</td></tr>
<tr><td colspan="6">计　划　对　比</td></tr>
<tr><td>序号</td><td>计划的可行性</td><td>计划的经济性</td><td>计划的可操作性</td><td>计划的实施难度</td><td>综合评价</td></tr>
<tr><td>方案 1</td><td></td><td></td><td></td><td></td><td></td></tr>
<tr><td>方案 2</td><td></td><td></td><td></td><td></td><td></td></tr>
<tr><td>方案 3</td><td></td><td></td><td></td><td></td><td></td></tr>
<tr><td rowspan="3">决策评价</td><td>班级</td><td></td><td>第（　　）组</td><td>组长签字</td><td></td></tr>
<tr><td>教师签字</td><td></td><td>日期</td><td colspan="2"></td></tr>
<tr><td colspan="5">评语：</td></tr>
</table>

3.1.7.4 攻螺纹环节实施单

<table>
<tr><td colspan="2">学习场</td><td colspan="4">金 工 实 训</td></tr>
<tr><td colspan="2">学习情境 3</td><td colspan="2">变压器低压套管线夹 M10 六角螺母的加工</td><td>本环节学时</td><td>20 分钟</td></tr>
<tr><td colspan="2">典型工作过程描述</td><td colspan="4">下料—锯割—锉端面—划端面正六边形—锉正六棱面—钻孔—攻螺纹—倒角</td></tr>
<tr><td>序号</td><td>实施步骤</td><td>工作内容及方法</td><td colspan="2">工艺质量要求</td><td>注意事项</td></tr>
<tr><td>1</td><td>工作前准备</td><td></td><td colspan="2"></td><td></td></tr>
<tr><td>2</td><td>开班前会，布置任务</td><td></td><td colspan="2"></td><td></td></tr>
<tr><td>3</td><td>安全操作攻螺纹</td><td>用 M10 的丝锥（一锥、二锥）攻螺纹，在攻螺纹过程中，两手用力要均衡，旋转要平衡，每旋转 1/2～1 周时，将丝锥反转 1/4 周，以排除切屑，防止切屑堵塞屑槽，造成丝锥损坏或者折断。</td><td colspan="2">螺纹顺滑。</td><td>锋利物伤害。</td></tr>
<tr><td>4</td><td>清理现场</td><td></td><td colspan="2"></td><td></td></tr>
<tr><td colspan="6">实施说明：</td></tr>
</table>

<table>
<tr><td rowspan="3">实施评价</td><td>班级</td><td></td><td>第（　　）组</td><td>组长签字</td><td></td></tr>
<tr><td>教师签字</td><td></td><td>日期</td><td colspan="2"></td></tr>
<tr><td colspan="5">评语：</td></tr>
</table>

3.1.7.5　攻螺纹环节检查单

学习场	金　工　实　训				
学习情境 3	变压器低压套管线夹 M10 六角螺母的加工			本环节学时	6 分钟
典型工作过程描述	下料—锯割—锉端面—划端面正六边形—锉正六棱面—钻孔—攻螺纹—倒角				
序号	检查项目	检查标准		学生自查	教师检查
1	资讯环节	获得 M10 六角螺母攻螺纹操作的有关信息的来源。			
2	计划环节	制定攻螺纹操作步骤的各个方案。			
3	决策环节	对各种攻螺纹操作的方案衡量对比评分，获取最佳方案。			
4	实施环节	攻螺纹操作过程中各个环节按照标准执行，贯彻落实“四不伤害”。			
5	检查环节	各个环节逐一检查，是否有爱岗敬业的主题思想。			
检查评价	班级		第（　　）组	组长签字	
	教师签字		日期		
	评语：				

学习情境 3-39

3.1.8 倒角

3.1.8.1 倒角环节资讯单

学习场	金 工 实 训		
学习情境 3	变压器低压套管线夹 M10 六角螺母的加工	本环节学时	6 分钟
典型工作过程描述	下料—锯割—锉端面—划端面正六边形—锉正六棱面—钻孔—攻螺纹—倒角		
搜集资讯的方式	教师的讲解、互联网查询或者同学间交流		
资讯描述	查看教师提供的资料，获取六角螺母倒角操作的有关信息，便于安全有效地完成任务。		
对学生的要求	课前充分预习实训作业单，充分利用网络查找有关实训的学习资料，实训过程中穿着实训服装，贯彻落实自己不伤害自己、自己不伤害他人、自己不被他人伤害、保护他人不被伤害的“四不伤害”安全要求，注意严格遵守实训室的各项规章制度。		
参考资料	参考《1 型六角螺母》(GB/T 6170—2015)。		

技 能 要 点

1. 下列工具中禁止用小锤敲打的是（　　）。

A. 划规　B. 样冲　C. 划针　D. 游标卡尺

2. 下列说法正确的是（　　）。

A. 锯条安装在锯弓上越紧越好

B. 锯割工件用虎钳夹持时，锯切位置不宜伸出过长

C. 工件锯割开始或者将要切断时，须轻轻推割，以防滑出伤手或使锯条断裂

D. 锯切工件一定夹紧，锯切钢件时要润滑

E. 使用锉刀应装上手柄

F. 锉刀放置不能伸出工作台外

G. 不准用锉刀撬、砸、敲打其他物品

3.1.8.2　倒角环节计划单

<table>
<tr><td>学习场</td><td colspan="5">金　工　实　训</td></tr>
<tr><td>学习情境 3</td><td colspan="3">变压器低压套管线夹 M10 六角螺母的加工</td><td>本环节学时</td><td>6 分钟</td></tr>
<tr><td>典型工作过程描述</td><td colspan="5">下料—锯割—锉端面—划端面正六边形—锉正六棱面—钻孔—攻螺纹—倒角</td></tr>
<tr><td>计划制订的方式</td><td colspan="5">同学之间小组讨论</td></tr>
<tr><td>序号</td><td colspan="2">工 作 步 骤</td><td colspan="3">注 意 事 项</td></tr>
<tr><td>1</td><td colspan="2">工作前准备</td><td colspan="3"></td></tr>
<tr><td>2</td><td colspan="2">开班前会，布置任务</td><td colspan="3"></td></tr>
<tr><td>3</td><td colspan="2">安全操作倒角</td><td colspan="3"></td></tr>
<tr><td>4</td><td colspan="2">清理现场</td><td colspan="3"></td></tr>
<tr><td rowspan="3">计划评价</td><td>班级</td><td></td><td>第（　　）组</td><td>组长签字</td><td></td></tr>
<tr><td>教师签字</td><td></td><td>日期</td><td colspan="2"></td></tr>
<tr><td colspan="5">评语：</td></tr>
</table>

3.1.8.3 倒角环节决策单

<table>
<tr><td>学习场</td><td colspan="5">金 工 实 训</td></tr>
<tr><td>学习情境 3</td><td colspan="3">变压器低压套管线夹 M10 六角螺母的加工</td><td>本环节学时</td><td>6 分钟</td></tr>
<tr><td>典型工作过程描述</td><td colspan="5">下料—锯割—锉端面—划端面正六边形—锉正六棱面—钻孔—攻螺纹—倒角</td></tr>
<tr><td colspan="6">计 划 对 比</td></tr>
<tr><td>序号</td><td>计划的可行性</td><td>计划的经济性</td><td>计划的可操作性</td><td>计划的实施难度</td><td>综合评价</td></tr>
<tr><td>方案 1</td><td></td><td></td><td></td><td></td><td></td></tr>
<tr><td>方案 2</td><td></td><td></td><td></td><td></td><td></td></tr>
<tr><td>方案 3</td><td></td><td></td><td></td><td></td><td></td></tr>
</table>

<table>
<tr><td rowspan="3">决策评价</td><td>班级</td><td></td><td>第（　　）组</td><td>组长签字</td><td></td></tr>
<tr><td>教师签字</td><td></td><td>日期</td><td colspan="2"></td></tr>
<tr><td colspan="5">评语：</td></tr>
</table>

3.1.8.4 倒角环节实施单

<table>
<tr><td colspan="2">学习场</td><td colspan="3">金 工 实 训</td></tr>
<tr><td colspan="2">学习情境 3</td><td>变压器低压套管线夹 M10 六角螺母的加工</td><td>本环节学时</td><td>30 分钟</td></tr>
<tr><td colspan="2">典型工作过程描述</td><td colspan="3">下料—锯割—锉端面—划端面正六边形—锉正六棱面—钻孔—攻螺纹—倒角</td></tr>
<tr><td>序号</td><td>实施步骤</td><td>工作内容及方法</td><td>工艺质量要求</td><td>注意事项</td></tr>
<tr><td>1</td><td>工作前准备</td><td></td><td></td><td></td></tr>
<tr><td>2</td><td>开班前会，布置任务</td><td></td><td></td><td></td></tr>
<tr><td>3</td><td>安全操作倒角</td><td>锉掉两个端面的六个角。</td><td>要求锉掉的各个角形成等边三角形。</td><td>锋利物伤害。</td></tr>
<tr><td>4</td><td>清理现场</td><td>1. 现场工作结束后，工作负责人应关闭电源，检查工具。
2. 清扫现场。</td><td>按照要求进行安全检查和工具验收，清扫现场。</td><td>铁屑伤害和锋利物伤害。</td></tr>
<tr><td colspan="5">实施说明：</td></tr>
</table>

<table>
<tr><td rowspan="3">实施评价</td><td>班级</td><td></td><td>第（　　）组</td><td>组长签字</td><td></td></tr>
<tr><td>教师签字</td><td></td><td>日期</td><td colspan="2"></td></tr>
<tr><td colspan="5">评语：</td></tr>
</table>

学习情境 3-43

3.1.8.5 倒角环节检查单

<table>
<tr><td colspan="2">学习场</td><td colspan="4">金 工 实 训</td></tr>
<tr><td colspan="2">学习情境 3</td><td colspan="2">变压器低压套管线夹 M10 六角螺母的加工</td><td>本环节学时</td><td>6 分钟</td></tr>
<tr><td colspan="2">典型工作过程描述</td><td colspan="4">下料—锯割—锉端面—划端面正六边形—锉正六棱面—钻孔—攻螺纹—倒角</td></tr>
<tr><td>序号</td><td>检查项目</td><td colspan="2">检 查 标 准</td><td>学生自查</td><td>教师检查</td></tr>
<tr><td>1</td><td>资讯环节</td><td colspan="2">获得 M10 六角螺母倒角操作的有关信息的来源。</td><td></td><td></td></tr>
<tr><td>2</td><td>计划环节</td><td colspan="2">制定倒角操作步骤的各个方案。</td><td></td><td></td></tr>
<tr><td>3</td><td>决策环节</td><td colspan="2">对各种倒角操作的方案衡量对比评分，获取最佳方案。</td><td></td><td></td></tr>
<tr><td>4</td><td>实施环节</td><td colspan="2">倒角操作过程中各个环节按照标准执行，贯彻落实“四不伤害”。</td><td></td><td></td></tr>
<tr><td>5</td><td>检查环节</td><td colspan="2">各个环节逐一检查，是否有爱岗敬业的主题思想。</td><td></td><td></td></tr>
<tr><td colspan="2" rowspan="3">检查评价</td><td>班级</td><td></td><td>第（ ）组</td><td>组长签字</td></tr>
<tr><td>教师签字</td><td></td><td>日期</td><td></td></tr>
<tr><td colspan="4">评语：</td></tr>
</table>

学习情境 3-44

3.2　变压器低压套管线夹 M10 六角螺母的加工学习情境材料工具清单

<table>
<tr><td>学习场</td><td colspan="3">金　工　实　训</td></tr>
<tr><td>学习情境 3</td><td>变压器低压套管线夹 M10 六角螺母的加工</td><td>本环节学时</td><td>540 分钟</td></tr>
<tr><td>典型工作过程描述</td><td colspan="3">下料—锯割—锉端面—划端面正六边形—锉正六棱面—钻孔—攻螺纹—倒角</td></tr>
</table>

序号	名称	作用	数量	型号	使用者

班级		第（　　）组	组长签字	
教师签字		日期		

3.3　变压器低压套管线夹 M10 六角螺母的加工学习情境教学反馈单

<table>
<tr><td>学习场</td><td colspan="5">金　工　实　训</td></tr>
<tr><td>学习情境 3</td><td colspan="3">变压器低压套管线夹 M10 六角螺母的加工</td><td>本环节学时</td><td>540 分钟</td></tr>
<tr><td>典型工作过程描述</td><td colspan="5">下料—锯割—锉端面—划端面正六边形—锉正六棱面—钻孔—攻螺纹—倒角</td></tr>
<tr><td rowspan="6">调查项目</td><td>序号</td><td colspan="2">调 查 内 容</td><td colspan="2">理 由 描 述</td></tr>
<tr><td></td><td colspan="2"></td><td colspan="2"></td></tr>
<tr><td></td><td colspan="2"></td><td colspan="2"></td></tr>
<tr><td></td><td colspan="2"></td><td colspan="2"></td></tr>
<tr><td></td><td colspan="2"></td><td colspan="2"></td></tr>
<tr><td></td><td colspan="2"></td><td colspan="2"></td></tr>
<tr><td colspan="6">您对本次课程教学的改进意见：</td></tr>
<tr><td>调查信息</td><td>被调查人姓名</td><td></td><td>调查日期</td><td colspan="2"></td></tr>
</table>

3.4　变压器低压套管线夹 M10 六角螺母的加工学习情境分组单

学习场	金　工　实　训					
学习情境 3	变压器低压套管线夹 M10 六角螺母的加工				本环节学时	540 分钟
典型工作过程描述	下料—锯割—锉端面—划端面正六边形—锉正六棱面—钻孔—攻螺纹—倒角					
分组情况	组别	组长	组　员			
分组说明						
班级		教师签字			日期	

3.5　变压器低压套管线夹M10六角螺母的加工学习情境评价单

<table>
<tr><td>学习场</td><td colspan="5">金　工　实　训</td></tr>
<tr><td>学习情境3</td><td colspan="2">变压器低压套管线夹M10六角螺母的加工</td><td colspan="2">本环节学时</td><td>20分钟</td></tr>
<tr><td>典型工作过程描述</td><td colspan="5">下料—锯割—锉端面—划端面正六边形—锉正六棱面—钻孔—攻螺纹—倒角</td></tr>
<tr><td>评价项目</td><td colspan="3">评 价 子 项 目</td><td>学生自评
20%</td><td>组内评价
30%</td><td>教师评价
50%</td></tr>
<tr><td>典型工作环节下料</td><td colspan="3">1. 资讯10分；2. 计划10分；3. 决策10分；4. 实施40分；5. 检查10分；6. 工作态度20分。</td><td></td><td></td><td></td></tr>
<tr><td>典型工作环节锯割</td><td colspan="3">1. 资讯10分；2. 计划10分；3. 决策10分；4. 实施40分；5. 检查10分；6. 工作态度20分。</td><td></td><td></td><td></td></tr>
<tr><td>典型工作环节
锉端面</td><td colspan="3">1. 资讯10分；2. 计划10分；3. 决策10分；4. 实施40分；5. 检查10分；6. 工作态度20分。</td><td></td><td></td><td></td></tr>
<tr><td>典型工作环节
划端面正六边形</td><td colspan="3">1. 资讯10分；2. 计划10分；3. 决策10分；4. 实施40分；5. 检查10分；6. 工作态度20分。</td><td></td><td></td><td></td></tr>
<tr><td>典型工作环节
锉正六棱面</td><td colspan="3">1. 资讯10分；2. 计划10分；3. 决策10分；4. 实施40分；5. 检查10分；6. 工作态度20分。</td><td></td><td></td><td></td></tr>
<tr><td>典型工作环节钻孔</td><td colspan="3">1. 资讯10分；2. 计划10分；3. 决策10分；4. 实施40分；5. 检查10分；6. 工作态度20分。</td><td></td><td></td><td></td></tr>
<tr><td>典型工作环节
攻螺纹</td><td colspan="3">1. 资讯10分；2. 计划10分；3. 决策10分；4. 实施40分；5. 检查10分；6. 工作态度20分。</td><td></td><td></td><td></td></tr>
<tr><td>典型工作环节倒角</td><td colspan="3">1. 资讯10分；2. 计划10分；3. 决策10分；4. 实施40分；5. 检查10分；6. 工作态度20分。</td><td></td><td></td><td></td></tr>
<tr><td>最终结果</td><td colspan="3">1. 资讯10分；2. 计划10分；3. 决策10分；4. 实施40分；5. 检查10分；6. 工作态度20分。</td><td></td><td></td><td></td></tr>
<tr><td rowspan="3">评价</td><td>班级</td><td></td><td>第（　　）组</td><td>组长签字</td><td colspan="2"></td></tr>
<tr><td>教师签字</td><td></td><td>日期</td><td colspan="3"></td></tr>
<tr><td colspan="6">评语：</td></tr>
</table>

3.6　教师实施计划单

学习场	金　工　实　训		
学习情境 3	变压器低压套管线夹 M10 六角螺母的加工	学时	540 分钟
典型工作过程描述	下料—锯割—锉端面—划端面正六边形—锉正六棱面—钻孔—攻螺纹—倒角		

序号	工作与学习步骤	学时	使用工具	地点	方式	备注

班级		教师签字		日期	

Learning scenario 3 Processing the M10 hexagon nut of transformer low-voltage bushing clamp

3.0 Learning worksheet

<table>
<tr><td>Learning field</td><td colspan="6">Metalworking practical training</td></tr>
<tr><td>Learning scenario 3</td><td colspan="3">Processing the M10 hexagon nut of transformer low-voltage bushing clamp</td><td colspan="2">Total teaching time</td><td>540 minutes</td></tr>
<tr><td>Description of the typical working processes</td><td colspan="6">Cutting stock – sawing – filing the end faces – scribing regular hexagon on end faces – filing the hexagonal side faces – drilling a hole – tapping – chamfering</td></tr>
<tr><td>Learning objectives</td><td colspan="6">1. Be able to use common tools, measuring tools correctly, and complete the simple processing of parts independently.
2. Be able to complete the processing of the workpiece, including procedures of scribe marking, sawing, and filing.
3. Be able to operate bench drilling safety and normatively.</td></tr>
<tr><td>Task description</td><td colspan="6">Process a piece of ϕ25mm round steel into a M10 hexagon nut after going through the procedures of cutting stock – filing the A end face – sawing – filing the B end faces – scribing regular hexagon on end faces – filing the hexagonal side faces – drilling a hole – tapping – chamfering – trimming. m = 8.4mm, S = 16mm, e = 17.77mm, as indicated in Figure 3.0.1.
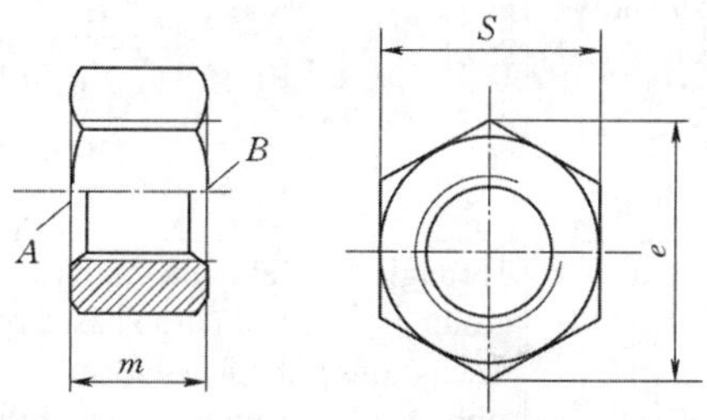

(a) Side face (b) End face
Figure 3.0.1 Hexagon nut</td></tr>
<tr><td>Teaching schedule</td><td>Information: 50 minutes</td><td>Planning: 50 minutes</td><td>Decision:50 minutes</td><td>Implementation: 320 minutes</td><td>Check: 50 minutes</td><td>Assessment: 20 minutes</td></tr>
<tr><td>Requirements for students</td><td colspan="6">Preview the practical training homework sheets before taking the class and use the Internet to find the learning materials related to our practical training. Wear the training uniform throughout the training process. Implement the "4 No-harm Principles" (do not harm yourself, do not harm others, do not get harmed by others, protect others from being harmed) and other safety requirements. Strictly follow the rules and regulations of the training room.</td></tr>
<tr><td>References</td><td colspan="6">Refer to Technical Code for Safety of Temporary Electrification on Construction Site (JGJ 46—2005) and Hexagon Nuts, Style 1 (GB/T 6170—2015).</td></tr>
</table>

3.1 Teaching guidance

Learning field	Metalworking practical training	Learning scenario 3	Processing the M10 hexagon nut of transformer low-voltage bushing clamp	Reference frame	*Hexagon Nuts, Style 1* (GB/T 6170—2015)	
General working processes / Typical working processes	Information	Planning	Decision	Implementation	Check	Assessment
Cutting stock	Check the *Hexagon Nuts, Style 1* (GB/T 6170—2015) to obtain relevant information about the M10 hexagon nut, so as to complete the tasks safely and effectively.	Through group discussions among students, prepare an electricity safety code scheme and complete 3.1.1.2 Planning sheet for cutting stock.	Take integrated evaluation based on the feasibility, economy, operability, and implementation difficulty of the plans, and then select the best one.	Check relevant information and complete 3.1.1.4 Implementation sheet for cutting stock.	Conduct self-check by students and check the by the teacher according to the check items and check criteria included in 3.1.1.5 Check sheet for cutting stock.	Conduct self-assessment, peer assessment, and teacher assessment according to the points assigned to the evaluation items, and complete 3.5 Assessment sheet for processing the M10 hexagon nut of transformer low-voltage bushing clamp learning scenario.
Sawing	Check the *Hexagon Nuts, Style 1* (GB/T 6170—2015) to obtain relevant information about the M10 hexagon nut, so as to complete the tasks safely and effectively.	Through group discussions among students, prepare an electricity safety code scheme and complete 3.1.2.2 Planning sheet for sawing.	Take integrated evaluation based on the feasibility, economy, operability, and implementation difficulty of the plans, and then select the best one.	Check relevant information and complete 3.1.2.4 Implementation sheet for sawing.	Conduct self-check by students and check by the teacher according to the check items and check criteria included in 3.1.2.5 Check sheet for sawing.	Conduct self-assessment, peer assessment, and teacher assessment according to the points assigned to the evaluation items, and complete 3.5 Assessment sheet for processing the M10 hexagon nut of transformer low-voltage bushing clamp learning scenario.

Continue

Learning field	Metalworking practical training	Learning scenario 3	Processing the M10 hexagon nut of transformer low-voltage bushing clamp	Reference frame	*Hexagon Nuts, Style 1* (GB/T 6170—2015)	
General working processes / Typical working processes	Information	Planning	Decision	Implementation	Check	Assessment
Filing the end faces	Check the *Hexagon Nuts, Style 1* (GB/T 6170—2015) to obtain relevant information about the M10 hexagon nut, so as to complete the tasks safely and effectively.	Through group discussions among students, prepare an electricity safety code scheme and complete 3.1.3.2 Planning sheet for filing the end faces	Take integrated evaluation based on the feasibility, economy, operability, and implementation difficulty of the plans, and then select the best one.	Check relevant information and complete 3.1.3.4 Implementation sheet for filing the end faces.	Conduct self-check by students and check by the teacher according to the check items and check criteria included in 3.1.3.5 Check sheet for filing the end faces.	Conduct self-assessment, peer assessment, and teacher assessment according to the points assigned to the evaluation items, and complete 3.5 Assessment sheet for processing the M10 hexagon nut of transformer low-voltage bushing clamp learning scenario.
Scribing regular hexagon on end faces	Check the *Hexagon Nuts, Style 1* (GB/T 6170—2015) to obtain relevant information about the M10 hexagon nut, so as to complete the tasks safely and effectively.	Through group discussions among students, prepare an electricity safety code scheme and complete 3.1.4.2 Planning sheet for scribing regular hexagon on end faces.	Take integrated evaluation based on the feasibility, economy, operability, and implementation difficulty of the plans, and then select the best one.	Check relevant information and complete 3.1.4.4 Implementation sheet for scribing regular hexagon on end faces.	Conduct self-check by students and check by the teacher according to the check items and check criteria included in 3.1.4.5 Check sheet for scribing regular hexagon on end faces.	Conduct self-assessment, peer assessment, and teacher assessment according to the points assigned to the evaluation items, and complete 3.5 Assessment sheet for processing the M10 hexagon nut of transformer low-voltage bushing clamp learning scenario.

Continue

Learning field	Metalworking practical training	Learning scenario 3	Processing the M10 hexagon nut of transformer low-voltage bushing clamp	Reference frame	*Hexagon Nuts, Style 1* (GB/T 6170—2015)	
General working processes / Typical working processes	Information	Planning	Decision	Implementation	Check	Assessment
Filing the hexagonal side faces	Check the *Hexagon Nuts, Style 1* (GB/T 6170—2015) to obtain relevant information about the M10 hexagon nut, so as to complete the tasks safely and effectively.	Through group discussions among students, prepare an electricity safety code scheme and complete 3.1.5.2 Planning sheet for filing the hexagonal side faces.	Take integrated evaluation based on the feasibility, economy, operability, and implementation difficulty of the plans, and then select the best one.	Check relevant information and complete 3.1.5.4 Implementation sheet for filing the hexagonal side faces.	Conduct self-check by students and check by the teacher according to the check items and check criteria included in 3.1.5.5 Check sheet for filing the hexagonal side faces.	Conduct self-assessment, peer assessment, and teacher assessment according to the points assigned to the evaluation items, and complete 3.5 Assessment sheet for processing the M10 hexagon nut of transformer low-voltage bushing clamp learning scenario.
Drilling a hole	Check the *Hexagon Nuts, Style 1* (GB/T 6170—2015) to obtain relevant information about the M10 hexagon nut, so as to complete the tasks safely and effectively.	Through group discussions among students, prepare an electricity safety code scheme and complete 3.1.6.2 Planning sheet for drilling hole.	Take integrated evaluation based on the feasibility, economy, operability, and implementation difficulty of the plans, and then select the best one.	Check relevant information and complete 3.1.6.4 Implementation sheet for drilling hole.	Conduct self-check by students and check by the teacher according to the check items and check criteria included in 3.1.6.5 Check sheet for drilling hole.	Conduct self-assessment, peer assessment, and teacher assessment according to the points assigned to the evaluation items, and complete 3.5 Assessment sheet for processing the M10 hexagon nut of transformer low-voltage bushing clamp learning scenario.

Continue

Learning field	Metalworking practical training	Learning scenario 3	Processing the M10 hexagon nut of transformer low-voltage bushing clamp	Reference frame	*Hexagon Nuts, Style 1* (GB/T 6170—2015)	
General working processes / Typical working processes	Information	Planning	Decision	Implementation	Check	Assessment
Tapping	Check the *Hexagon Nuts, Style 1* (GB/T 6170—2015) to obtain relevant information about the M10 hexagon nut, so as to complete the tasks safely and effectively.	Through group discussions among students, prepare an electricity safety code scheme and complete 3.1.7.2 Planning sheet for tapping.	Take integrated evaluation based on the feasibility, economy, operability, and implementation difficulty of the plans, and then select the best one.	Check relevant information and complete 3.1.7.4 Implementation sheet for tapping.	Conduct self-check by students and check by the teacher according to the check items and check criteria included in 3.1.7.5 Check sheet for tapping.	Conduct self-assessment, peer assessment, and teacher assessment according to the points assigned to the evaluation items, and complete 3.5 Assessment sheet for processing the M10 hexagon nut of transformer low-voltage bushing clamp learning scenario.
Chamfering	Check the *Hexagon Nuts, Style 1* (GB/T 6170—2015) to obtain relevant information about the M10 hexagon nut, so as to complete the tasks safely and effectively.	Through group discussions among students, prepare an electricity safety code scheme and complete 3.1.8.2 Planning sheet for chamfering.	Take integrated evaluation based on the feasibility, economy, operability, and implementation difficulty of the plans, and then select the best one.	Check relevant information and complete 3.1.8.4 Implementation sheet for chamfering.	Conduct self-check by students and check by the teacher according to the check items and check criteria included in 3.1.8.5 Check sheet for chamfering.	Conduct self-assessment, peer assessment, and teacher assessment according to the points assigned to the evaluation items, and complete 3.5 Assessment sheet for processing the M10 hexagon nut of transformer low-voltage bushing clamp learning scenario.

3.1.1 Cutting stock

3.1.1.1 Information sheet for cutting stock

<table>
<tr><td>Learning field</td><td colspan="3">Metalworking practical training</td></tr>
<tr><td>Learning scenario 3</td><td>Processing the M10 hexagon nut of transformer low-voltage bushing clamp</td><td>Session teaching time</td><td>6 minutes</td></tr>
<tr><td>Description of the typical working processes</td><td colspan="3">Cutting stock – sawing – filing the end faces – scribing regular hexagon on end faces – filing the hexagonal side faces – drilling a hole – tapping – chamfering</td></tr>
<tr><td>Information collection methods</td><td colspan="3">Teacher's explanation, search on the Internet, or communication among students</td></tr>
<tr><td>Information description</td><td colspan="3">Review the materials provided by the instructor to obtain information about cutting stock of the M10 hexagon nut, so as to complete the tasks safely and effectively, as indicated in Figure 3.1.1.

Figure 3.1.1 The M10 hexagon nut of the wire clamp</td></tr>
<tr><td>Requirements for students</td><td colspan="3">Preview the practical training homework sheets before taking the class and use the Internet to find the learning materials related to our practical training. Wear the training uniform throughout the training process. Implement the "4 No-harm Principles" (do not harm yourself, do not harm others, do not get harmed by others, protect others from being harmed) and other safety requirements. Strictly follow the rules and regulations of the training room.</td></tr>
<tr><td>References</td><td colspan="3">Refer to Technical Code for Safety of Temporary Electrification on Construction Site (JGJ 46—2005) and Hexagon Nuts, Style 1 (GB/T 6170—2015).</td></tr>
</table>

3.1.1.2 Planning sheet for cutting stock

Learning field	Metalworking practical training		
Learning scenario 3	Processing the M10 hexagon nut of transformer low-voltage bushing clamp	Session teaching time	6 minutes
Description of the typical working processes	Cutting stock – sawing – filing the end faces – scribing regular hexagon on end faces – filing the hexagonal side faces – drilling a hole – tapping – chamfering		
Planning methods	Group discussion among classmates		

Number	Working steps	Notice
1	Preparation before work	
2	Conduct a pre-class meeting, allocate the tasks	
3	Safe operation	
4	Acceptance of products	

<table>
<tr><td rowspan="3">Plan assessment</td><td>Class</td><td></td><td>Group ( )</td><td>Group leader's signature</td><td></td></tr>
<tr><td>Teacher's signature</td><td></td><td>Date</td><td colspan="2"></td></tr>
<tr><td colspan="5">Comments:</td></tr>
</table>

3.1.1.3 Decision sheet for cutting stock

<table>
<tr><td>Learning field</td><td colspan="5">Metalworking practical training</td></tr>
<tr><td>Learning scenario 3</td><td colspan="3">Processing the M10 hexagon nut of transformer low-voltage bushing clamp</td><td>Session teaching time</td><td>6 minutes</td></tr>
<tr><td>Description of the typical working processes</td><td colspan="5">Cutting stock – sawing – filing the end faces – scribing regular hexagon on end faces – filing the hexagonal side faces – drilling a hole – tapping – chamfering</td></tr>
<tr><td colspan="6">Plan comparison</td></tr>
<tr><td>Number</td><td>Feasibility</td><td>Economy</td><td>Operability</td><td>Implementation difficulty</td><td>Integrated evaluation</td></tr>
<tr><td>Plan 1</td><td></td><td></td><td></td><td></td><td></td></tr>
<tr><td>Plan 2</td><td></td><td></td><td></td><td></td><td></td></tr>
<tr><td>Plan 3</td><td></td><td></td><td></td><td></td><td></td></tr>
<tr><td rowspan="3">Decision assessment</td><td>Class</td><td></td><td>Group ( )</td><td>Group leader's signature</td><td></td></tr>
<tr><td>Teacher's signature</td><td></td><td>Date</td><td colspan="2"></td></tr>
<tr><td colspan="5">Comments:</td></tr>
</table>

3.1.1.4 Implementation sheet for cutting stock

<table>
<tr><td>Learning field</td><td colspan="4">Metalworking practical training</td></tr>
<tr><td>Learning scenario 3</td><td colspan="2">Processing the M10 hexagon nut of transformer low-voltage bushing clamp</td><td>Session teaching time</td><td>30 minutes</td></tr>
<tr><td>Description of the typical working processes</td><td colspan="4">Cutting stock – sawing – filing the end faces – scribing regular hexagon on end faces – filing the hexagonal side faces – drilling a hole – tapping – chamfering</td></tr>
<tr><td>Number</td><td>Implementation steps</td><td>Operations and methods</td><td>Process quality requirements</td><td>Notice</td></tr>
<tr><td>1</td><td>Preparation before work</td><td></td><td></td><td></td></tr>
<tr><td>2</td><td>Conduct a pre-class meeting, allocate the tasks</td><td></td><td></td><td></td></tr>
<tr><td>3</td><td>Safe operation</td><td>Use a ruler or vernier caliper to check the size of the imported materials – ϕ25mm round steel.</td><td>The length of the ϕ 25mm round steel is greater than 10mm and less than 15mm, and the end face is plumped without defects.</td><td>Avoid being harmed by iron filings and sharp objects.</td></tr>
<tr><td>4</td><td>Acceptance of products</td><td></td><td></td><td></td></tr>
<tr><td colspan="5">Implementation instructions:</td></tr>
</table>

<table>
<tr><td rowspan="3">Implementation assessment</td><td>Class</td><td></td><td>Group ( )</td><td>Group leader's signature</td><td></td></tr>
<tr><td>Teacher's signature</td><td></td><td>Date</td><td colspan="2"></td></tr>
<tr><td colspan="5">Comments:</td></tr>
</table>

3.1.1.5 Check sheet for cutting stock

<table>
<tr><td>Learning field</td><td colspan="5">Metalworking practical training</td></tr>
<tr><td>Learning scenario 3</td><td colspan="3">Processing the M10 hexagon nut of transformer low-voltage bushing clamp</td><td>Session teaching time</td><td>6 minutes</td></tr>
<tr><td>Description of the typical working processes</td><td colspan="5">Cutting stock – sawing – filing the end faces – scribing regular hexagon on end faces – filing the hexagonal side faces – drilling a hole – tapping – chamfering</td></tr>
<tr><td>Number</td><td>Check items</td><td colspan="2">Check criteria</td><td>Self-check by students</td><td>Check by the teacher</td></tr>
<tr><td>1</td><td>Cutting stock information</td><td colspan="2">Obtain the source of information about cutting stock of the M10 hexagon nut.</td><td></td><td></td></tr>
<tr><td>2</td><td>Cutting stock planning</td><td colspan="2">Formulate relevant plans for cutting stock.</td><td></td><td></td></tr>
<tr><td>3</td><td>Cutting stock decision</td><td colspan="2">Compare scores of various cutting stock plans and select the best solution.</td><td></td><td></td></tr>
<tr><td>4</td><td>Cutting stock implementation</td><td colspan="2">Operate each step of cutting stock in accordance with the standards and strictly implement the “4 No-harm Principles”.</td><td></td><td></td></tr>
<tr><td>5</td><td>Cutting stock check</td><td colspan="2">Check each step carefully, and check whether the theme idea of dedication to work is followed.</td><td></td><td></td></tr>
<tr><td rowspan="3">Check assessment</td><td>Class</td><td></td><td>Group ( )</td><td>Group leader’s signature</td><td></td></tr>
<tr><td>Teacher’s signature</td><td></td><td>Date</td><td colspan="2"></td></tr>
<tr><td colspan="5">Comments:</td></tr>
</table>

3.1.2 Sawing

3.1.2.1 Information sheet for sawing

<table>
<tr><td>Learning field</td><td colspan="3">Metalworking practical training</td></tr>
<tr><td>Learning scenario 3</td><td>Processing the M10 hexagon nut of transformer low-voltage bushing clamp</td><td>Session teaching time</td><td>6 minutes</td></tr>
<tr><td>Description of the typical working processes</td><td colspan="3">Cutting stock – sawing – filing the end faces – scribing regular hexagon on end faces – filing the hexagonal side faces – drilling a hole – tapping – chamfering</td></tr>
<tr><td>Information collection methods</td><td colspan="3">Teacher's explanation, search on the Internet, or communication among students</td></tr>
<tr><td>Information description</td><td colspan="3">Review the materials provided by the instructor to obtain information about sawing, so as to complete the tasks safely and effectively, as indicated in Figure 3.1.2.

Figure 3.1.2 On-site sawing</td></tr>
<tr><td>Requirements for students</td><td colspan="3">Preview the practical training homework sheets before taking the class and use the Internet to find the learning materials related to our practical training. Wear the training uniform throughout the training process. Implement the "4 No-harm Principles" (do not harm yourself, do not harm others, do not get harmed by others, protect others from being harmed) and other safety requirements. Strictly follow the rules and regulations of the training room.</td></tr>
<tr><td>References</td><td colspan="3">Refer to Hexagon Nuts, Style 1 (GB/T 6170—2015).
Operation of sawing</td></tr>
</table>

3.1.2.2 Planning sheet for sawing

<table>
<tr><td>Learning field</td><td colspan="5">Metalworking practical training</td></tr>
<tr><td>Learning scenario 3</td><td colspan="2">Processing the M10 hexagon nut of transformer low-voltage bushing clamp</td><td colspan="2">Session teaching time</td><td>6 minutes</td></tr>
<tr><td>Description of the typical working processes</td><td colspan="5">Cutting stock – sawing – filing the end faces – scribing regular hexagon on end faces – filing the hexagonal side faces – drilling a hole – tapping – chamfering</td></tr>
<tr><td>Planning methods</td><td colspan="5">Group discussion among classmates</td></tr>
<tr><td>Number</td><td>Working steps</td><td colspan="4">Safe and civilized production</td></tr>
<tr><td>1</td><td>Check the bow saw’s condition</td><td colspan="4"></td></tr>
<tr><td>2</td><td>Install the saw blade</td><td colspan="4"></td></tr>
<tr><td>3</td><td>Saw</td><td colspan="4"></td></tr>
<tr><td>4</td><td>Clean up the site</td><td colspan="4"></td></tr>
<tr><td rowspan="3">Plan assessment</td><td>Class</td><td></td><td>Group ( )</td><td>Group leader’s signature</td><td></td></tr>
<tr><td>Teacher’s signature</td><td></td><td>Date</td><td colspan="2"></td></tr>
<tr><td colspan="5">Comments:</td></tr>
</table>

3.1.2.3 Decision sheet for sawing

<table>
<tr><td>Learning field</td><td colspan="5">Metalworking practical training</td></tr>
<tr><td>Learning scenario 3</td><td colspan="3">Processing the M10 hexagon nut of transformer low-voltage bushing clamp</td><td>Session teaching time</td><td>6 minutes</td></tr>
<tr><td>Description of the typical working processes</td><td colspan="5">Cutting stock – sawing – filing the end faces – scribing regular hexagon on end faces – filing the hexagonal side faces – drilling a hole – tapping – chamfering</td></tr>
<tr><td colspan="6">Plan comparison</td></tr>
<tr><td>Number</td><td>Feasibility</td><td>Economy</td><td>Operability</td><td>Implementation difficulty</td><td>Integrated evaluation</td></tr>
<tr><td>Plan 1</td><td></td><td></td><td></td><td></td><td></td></tr>
<tr><td>Plan 2</td><td></td><td></td><td></td><td></td><td></td></tr>
<tr><td>Plan 3</td><td></td><td></td><td></td><td></td><td></td></tr>
<tr><td rowspan="3">Decision assessment</td><td>Class</td><td></td><td>Group ( )</td><td>Group leader's signature</td><td></td></tr>
<tr><td>Teacher's signature</td><td></td><td>Date</td><td colspan="2"></td></tr>
<tr><td colspan="5">Comments:</td></tr>
</table>

3.1.2.4 Implementation sheet for sawing

<table>
<tr><td>Learning field</td><td colspan="5">Metalworking practical training</td></tr>
<tr><td>Learning scenario 3</td><td colspan="2">Processing the M10 hexagon nut of transformer low-voltage bushing clamp</td><td colspan="2">Session teaching time</td><td>30 minutes</td></tr>
<tr><td>Description of the typical working processes</td><td colspan="5">Cutting stock – sawing – filing the end faces – scribing regular hexagon on end faces – filing the hexagonal side faces – drilling a hole – tapping – chamfering</td></tr>
<tr><td>Number</td><td>Implementation steps</td><td>Operations and methods</td><td colspan="2">Process quality requirements</td><td>Notice</td></tr>
<tr><td>1</td><td>Check the bow saw’s condition</td><td></td><td colspan="2"></td><td></td></tr>
<tr><td>2</td><td>Install the saw blade</td><td></td><td colspan="2"></td><td></td></tr>
<tr><td>3</td><td>Safe operation of sawing</td><td>Saw off a 10mm long piece of steel with the filed end face *A* [Figure 3.0.1 (a)] as the base plane.</td><td colspan="2">The opposite surface of the filed end face *A* [Figure 3.0.1 (a)] should be basically smooth and flat, and the length of the steel material should be 10mm.</td><td>Avoid being harmed by sawtooth.</td></tr>
<tr><td>4</td><td>Clean up the site</td><td></td><td colspan="2"></td><td></td></tr>
<tr><td colspan="6">Implementation instructions:</td></tr>
<tr><td rowspan="3">Implementation assessment</td><td>Class</td><td></td><td>Group ( )</td><td>Group leader’s signature</td><td></td></tr>
<tr><td>Teacher’s signature</td><td></td><td>Date</td><td colspan="2"></td></tr>
<tr><td colspan="5">Comments:</td></tr>
</table>

3.1.2.5 Check sheet for sawing

<table>
<tr><td>Learning field</td><td colspan="5">Metalworking practical training</td></tr>
<tr><td>Learning scenario 3</td><td colspan="2">Processing the M10 hexagon nut of transformer low-voltage bushing clamp</td><td colspan="2">Session teaching time</td><td>6 minutes</td></tr>
<tr><td>Description of the typical working processes</td><td colspan="5">Cutting stock – sawing – filing the end faces – scribing regular hexagon on end faces – filing the hexagonal side faces – drilling a hole – tapping – chamfering</td></tr>
<tr><td>Number</td><td>Check items</td><td>Check criteria</td><td colspan="2">Self-check by students</td><td>Check by the teacher</td></tr>
<tr><td>1</td><td>Sawing information</td><td>Obtain the source of information about sawing off the M10 hexagon nut.</td><td colspan="2"></td><td></td></tr>
<tr><td>2</td><td>Sawing planning</td><td>Formulate relevant plans for sawing.</td><td colspan="2"></td><td></td></tr>
<tr><td>3</td><td>Sawing decision</td><td>Compare scores of various sawing plans and select the best solution.</td><td colspan="2"></td><td></td></tr>
<tr><td>4</td><td>Sawing implementation</td><td>Operate each step of sawing in accordance with the standards and strictly implement the "4 No-harm Principles".</td><td colspan="2"></td><td></td></tr>
<tr><td>5</td><td>Sawing check</td><td>Check each step carefully, and check whether the theme idea of dedication to work is followed.</td><td colspan="2"></td><td></td></tr>
<tr><td rowspan="3">Check assessment</td><td>Class</td><td></td><td>Group ( )</td><td>Group leader's signature</td><td></td></tr>
<tr><td>Teacher's signature</td><td></td><td>Date</td><td colspan="2"></td></tr>
<tr><td colspan="5">Comments:</td></tr>
</table>

3.1.3 Filing the end faces

3.1.3.1 Information sheet for filing the end faces

<table>
<tr><td>Learning field</td><td colspan="3">Metalworking practical training</td></tr>
<tr><td>Learning scenario 3</td><td>Processing the M10 hexagon nut of transformer low-voltage bushing clamp</td><td>Session teaching time</td><td>6 minutes</td></tr>
<tr><td>Description of the typical working processes</td><td colspan="3">Cutting stock – sawing – filing the end faces – scribing regular hexagon on end faces – filing the hexagonal side faces – drilling a hole – tapping – chamfering</td></tr>
<tr><td>Information collection methods</td><td colspan="3">Teacher’s explanation, search on the Internet, or communication among students</td></tr>
<tr><td>Information description</td><td colspan="3">Review the materials provided by the instructor to obtain information about filing, so as to complete the tasks safely and effectively, as indicated in Figure 3.1.3.

Figure 3.1.3 On-site filing</td></tr>
<tr><td>Requirements for students</td><td colspan="3">Preview the practical training homework sheets before taking the class and use the Internet to find the learning materials related to our practical training. Wear the training uniform throughout the training process. Implement the “4 No-harm Principles” (do not harm yourself, do not harm others, do not get harmed by others, protect others from being harmed) and other safety requirements. Strictly follow the rules and regulations of the training room.</td></tr>
<tr><td>References</td><td colspan="3">Refer to Hexagon Nuts, Style 1 (GB/T 6170—2015).
Operation of filing</td></tr>
</table>

3.1.3.2 Planning sheet for filing the end faces

<table>
<tr><td>Learning field</td><td colspan="5">Metalworking practical training</td></tr>
<tr><td>Learning scenario 3</td><td colspan="3">Processing the M10 hexagon nut of transformer low-voltage bushing clamp</td><td>Session teaching time</td><td>6 minutes</td></tr>
<tr><td>Description of the typical working processes</td><td colspan="5">Cutting stock – sawing – filing the end faces – scribing regular hexagon on end faces – filing the hexagonal side faces – drilling a hole – tapping – chamfering</td></tr>
<tr><td>Planning methods</td><td colspan="5">Group discussion among classmates</td></tr>
<tr><td>Number</td><td colspan="2">Working steps</td><td colspan="3">Notice</td></tr>
<tr><td>1</td><td colspan="2">Preparation before work</td><td colspan="3"></td></tr>
<tr><td>2</td><td colspan="2">Conduct a pre-class meeting, allocate the tasks</td><td colspan="3"></td></tr>
<tr><td>3</td><td colspan="2">Safe operation</td><td colspan="3">Avoid being harmed by iron filings and sharp objects.</td></tr>
<tr><td>4</td><td colspan="2">Check whether the surface is smooth and flat by using the light leakage method</td><td colspan="3"></td></tr>
<tr><td>5</td><td colspan="2">Clean up the site</td><td colspan="3"></td></tr>
<tr><td rowspan="3">Plan assessment</td><td>Class</td><td></td><td>Group ( )</td><td>Group leader's signature</td><td></td></tr>
<tr><td>Teacher's signature</td><td></td><td>Date</td><td></td><td></td></tr>
<tr><td colspan="5">Comments:</td></tr>
</table>

3.1.3.3 Decision sheet for filing the end faces

<table>
<tr><td>Learning field</td><td colspan="5">Metalworking practical training</td></tr>
<tr><td>Learning scenario 3</td><td colspan="3">Processing the M10 hexagon nut of transformer low-voltage bushing clamp</td><td>Session teaching time</td><td>6 minutes</td></tr>
<tr><td>Description of the typical working processes</td><td colspan="5">Cutting stock – sawing – filing the end faces – scribing regular hexagon on end faces – filing the hexagonal side faces – drilling a hole – tapping – chamfering</td></tr>
<tr><td colspan="6">Plan comparison</td></tr>
<tr><td>Number</td><td>Feasibility</td><td>Economy</td><td>Operability</td><td>Implementation difficulty</td><td>Integrated evaluation</td></tr>
<tr><td>Plan 1</td><td></td><td></td><td></td><td></td><td></td></tr>
<tr><td>Plan 2</td><td></td><td></td><td></td><td></td><td></td></tr>
<tr><td>Plan 3</td><td></td><td></td><td></td><td></td><td></td></tr>
<tr><td rowspan="3">Decision assessment</td><td>Class</td><td></td><td>Group ( )</td><td>Group leader's signature</td><td></td></tr>
<tr><td>Teacher's signature</td><td></td><td>Date</td><td colspan="2"></td></tr>
<tr><td colspan="5">Comments:</td></tr>
</table>

3.1.3.4 Implementation sheet for filing the end faces

<table>
<tr><td colspan="2">Learning field</td><td colspan="3">Metalworking practical training</td></tr>
<tr><td colspan="2">Learning scenario 3</td><td>Processing the M10 hexagon nut of transformer low-voltage bushing clamp</td><td>Session teaching time</td><td>60 minutes</td></tr>
<tr><td colspan="2">Description of the typical working processes</td><td colspan="3">Cutting stock – sawing – filing the end faces – scribing regular hexagon on end faces – filing the hexagonal side faces – drilling a hole – tapping – chamfering</td></tr>
<tr><td>Number</td><td>Implementation steps</td><td>Operations and methods</td><td>Process quality requirements</td><td>Notice</td></tr>
<tr><td>1</td><td>Preparation before work</td><td></td><td></td><td></td></tr>
<tr><td>2</td><td>Conduct a pre-class meeting, allocate the tasks</td><td></td><td></td><td></td></tr>
<tr><td>3</td><td>Safe operation</td><td>With end face A [Figure 3.0.1 (a)] as the base face, file the opposite side roughly and finely, so as to achieve the requirements of dimensional tolerance, parallelism and surface roughness.</td><td>Use the angle square for the light leakage method to check the roughness of end face B [Figure 3.0.1 (a)], and the required height of the two end faces is 8.4mm.</td><td>Avoid being harmed by iron filings and sharp objects.</td></tr>
<tr><td>4</td><td>Check whether the surface is smooth and flat with the light leakage method</td><td></td><td></td><td></td></tr>
<tr><td>5</td><td>Clean up the site</td><td></td><td></td><td></td></tr>
<tr><td colspan="5">Implementation instructions:</td></tr>
</table>

<table>
<tr><td rowspan="3">Implementation assessment</td><td>Class</td><td></td><td>Group ( )</td><td>Group leader's signature</td><td></td></tr>
<tr><td>Teacher's signature</td><td></td><td>Date</td><td colspan="2"></td></tr>
<tr><td colspan="5">Comments:</td></tr>
</table>

3.1.3.5 Check sheet for filing the end faces

<table>
<tr><td>Learning field</td><td colspan="5">Metalworking practical training</td></tr>
<tr><td>Learning scenario 3</td><td colspan="2">Processing the M10 hexagon nut of transformer low-voltage bushing clamp</td><td colspan="2">Session teaching time</td><td>6 minutes</td></tr>
<tr><td>Description of the typical working processes</td><td colspan="5">Cutting stock – sawing – filing the end faces – scribing regular hexagon on end faces – filing the hexagonal side faces – drilling a hole – tapping – chamfering</td></tr>
<tr><td>Number</td><td>Check items</td><td>Check criteria</td><td>Self-check by students</td><td colspan="2">Check by the teacher</td></tr>
<tr><td>1</td><td>Information about filing the end faces</td><td>Obtain the source of information about filing the M10 hexagon nut.</td><td></td><td colspan="2"></td></tr>
<tr><td>2</td><td>Planning of filing the end faces</td><td>Formulate relevant plans for filing.</td><td></td><td colspan="2"></td></tr>
<tr><td>3</td><td>Decision of filing the end faces</td><td>Compare scores of various filing plans and select the best solution.</td><td></td><td colspan="2"></td></tr>
<tr><td>4</td><td>Implementation of filing the end faces</td><td>Operate each step of filing in accordance with the standards and strictly implement the “4 No-harm Principles”.</td><td></td><td colspan="2"></td></tr>
<tr><td>5</td><td>Check of filing the end faces</td><td>Check each step carefully, and check whether the theme idea of dedication to work is followed.</td><td></td><td colspan="2"></td></tr>
<tr><td rowspan="3">Check assessment</td><td>Class</td><td></td><td>Group ( )</td><td>Group leader’s signature</td><td></td></tr>
<tr><td>Teacher’s signature</td><td></td><td>Date</td><td colspan="2"></td></tr>
<tr><td colspan="5">Comments:</td></tr>
</table>

3.1.4 Scribing regular hexagon on end faces

3.1.4.1 Information sheet for scribing regular hexagon on end faces

Learning field	Metalworking practical training		
Learning scenario 3	Processing the M10 hexagon nut of transformer low-voltage bushing clamp	Session teaching time	6 minutes
Description of the typical working processes	Cutting stock – sawing – filing the end faces – scribing regular hexagon on end faces – filing the hexagonal side faces – drilling a hole – tapping – chamfering		
Information collection methods	Teacher's explanation, search on the Internet, or communication among students		
Information description	Review the materials provided by the instructor to obtain information about scribing on hexagon nut, so as to complete the tasks safely and effectively, as indicated in Figure 3.1.4. 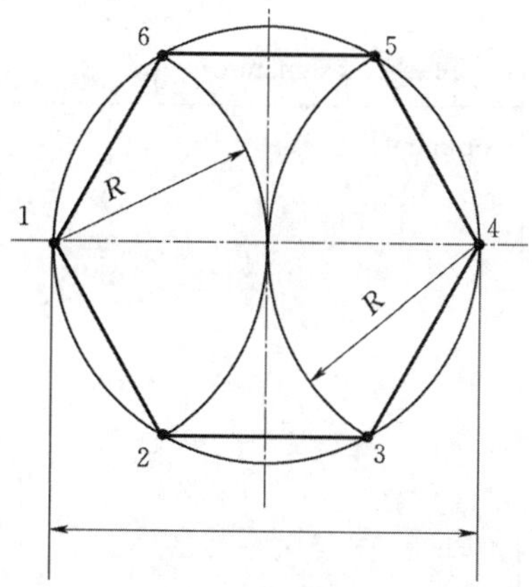 Figure 3.1.4 Scribing regular hexagon on end faces		
Requirements for students	Preview the practical training homework sheets before taking the class and use the Internet to find the learning materials related to our practical training. Wear the training uniform throughout the training process. Implement the "4 No-harm Principles" (do not harm yourself, do not harm others, do not get harmed by others, protect others from being harmed) and other safety requirements. Strictly follow the rules and regulations of the training room.		
References	Refer to *Hexagon Nuts, Style 1* (GB/T 6170—2015).		

3.1.4.2　Planning sheet for scribing regular hexagon on end faces

<table>
<tr><td>Learning field</td><td colspan="5">Metalworking practical training</td></tr>
<tr><td>Learning scenario 3</td><td colspan="3">Processing the M10 hexagon nut of transformer low-voltage bushing clamp</td><td>Session teaching time</td><td>6 minutes</td></tr>
<tr><td>Description of the typical working processes</td><td colspan="5">Cutting stock – sawing – filing the end faces – scribing regular hexagon on end faces – filing the hexagonal side faces – drilling a hole – tapping – chamfering</td></tr>
<tr><td>Planning methods</td><td colspan="5">Group discussion among classmates</td></tr>
<tr><td>Number</td><td>Working steps</td><td colspan="4">Notice</td></tr>
<tr><td>1</td><td>Preparation before work</td><td colspan="4"></td></tr>
<tr><td>2</td><td>Conduct a pre-class meeting, allocate the tasks</td><td colspan="4"></td></tr>
<tr><td>3</td><td>Safe operation of scribing</td><td colspan="4"></td></tr>
<tr><td>4</td><td>Clean up the site</td><td colspan="4"></td></tr>
<tr><td rowspan="3">Plan assessment</td><td>Class</td><td></td><td>Group ( )</td><td>Group leader's signature</td><td></td></tr>
<tr><td>Teacher's signature</td><td></td><td>Date</td><td colspan="2"></td></tr>
<tr><td colspan="5">Comments:</td></tr>
</table>

3.1.4.3 Decision sheet for scribing regular hexagon on end faces

<table>
<tr><td>Learning field</td><td colspan="5">Metalworking practical training</td></tr>
<tr><td>Learning scenario 3</td><td colspan="3">Processing the M10 hexagon nut of transformer low-voltage bushing clamp</td><td>Session teaching time</td><td>6 minutes</td></tr>
<tr><td>Description of the typical working processes</td><td colspan="5">Cutting stock – sawing – filing the end faces – scribing regular hexagon on end faces – filing the hexagonal side faces – drilling a hole – tapping – chamfering</td></tr>
<tr><td colspan="6">Plan comparison</td></tr>
<tr><td>Number</td><td>Feasibility</td><td>Economy</td><td>Operability</td><td>Implementation difficulty</td><td>Integrated evaluation</td></tr>
<tr><td>Plan 1</td><td></td><td></td><td></td><td></td><td></td></tr>
<tr><td>Plan 2</td><td></td><td></td><td></td><td></td><td></td></tr>
<tr><td>Plan 3</td><td></td><td></td><td></td><td></td><td></td></tr>
<tr><td rowspan="3">Decision assessment</td><td>Class</td><td></td><td>Group ( )</td><td>Group leader's signature</td><td></td></tr>
<tr><td>Teacher's signature</td><td></td><td>Date</td><td colspan="2"></td></tr>
<tr><td colspan="5">Comments:</td></tr>
</table>

3.1.4.4 Implementation sheet for scribing regular hexagon on end faces

<table>
<tr><td>Learning field</td><td colspan="4">Metalworking practical training</td></tr>
<tr><td>Learning scenario 3</td><td colspan="2">Processing the M10 hexagon nut of transformer low-voltage bushing clamp</td><td>Session teaching time</td><td>30 minutes</td></tr>
<tr><td>Description of the typical working processes</td><td colspan="4">Cutting stock – sawing – filing the end faces – scribing regular hexagon on end faces – filing the hexagonal side faces – drilling a hole – tapping – chamfering</td></tr>
<tr><td>Number</td><td>Implementation steps</td><td>Operations and methods</td><td>Process quality requirements</td><td>Notice</td></tr>
<tr><td>1</td><td>Preparation before work</td><td></td><td></td><td></td></tr>
<tr><td>2</td><td>Conduct a pre-class meeting, allocate the tasks</td><td></td><td></td><td></td></tr>
<tr><td>3</td><td>Safe operation of scribing</td><td>Use the compass, scriber, anvil, and caliper to scribe the regular hexagon processing boundary line of the hexagon nut.</td><td>The required distance between the opposite sides of the regular hexagon is 16mm.</td><td>Avoid being harmed by sharp objects.</td></tr>
<tr><td>4</td><td>Clean up the site</td><td></td><td></td><td></td></tr>
<tr><td colspan="5">Implementation instructions:</td></tr>
</table>

<table>
<tr><td rowspan="3">Implementation assessment</td><td>Class</td><td></td><td>Group ( )</td><td>Group leader's signature</td><td></td></tr>
<tr><td>Teacher's signature</td><td></td><td>Date</td><td colspan="2"></td></tr>
<tr><td colspan="5">Comments:</td></tr>
</table>

Without the rain, there would never be rainbows. Practice hard, then achieve remarkably.

3.1.4.5 Check sheet for scribing regular hexagon on end faces

<table>
<tr><td>Learning field</td><td colspan="4">Metalworking practical training</td></tr>
<tr><td>Learning scenario 3</td><td colspan="2">Processing the M10 hexagon nut of transformer low-voltage bushing clamp</td><td>Session teaching time</td><td>6 minutes</td></tr>
<tr><td>Description of the typical working processes</td><td colspan="4">Cutting stock – sawing – filing the end faces – scribing regular hexagon on end faces – filing the hexagonal side faces – drilling a hole – tapping – chamfering</td></tr>
<tr><td>Number</td><td>Check items</td><td>Check criteria</td><td>Self-check by students</td><td>Check by the teacher</td></tr>
<tr><td>1</td><td>Information</td><td>Obtain the source of information about scribing in the processing of the M10 hexagon nut.</td><td></td><td></td></tr>
<tr><td>2</td><td>Planning</td><td>Formulate relevant plans for scribing.</td><td></td><td></td></tr>
<tr><td>3</td><td>Decision</td><td>Compare scores of various scribing plans and select the best solution.</td><td></td><td></td></tr>
<tr><td>4</td><td>Implementation</td><td>Operate each step of scribing in accordance with the standards and strictly implement the “4 No-harm Principles”.</td><td></td><td></td></tr>
<tr><td>5</td><td>Check</td><td>Check each step carefully, and check whether the theme idea of dedication to work is followed.</td><td></td><td></td></tr>
</table>

<table>
<tr><td rowspan="3">Check assessment</td><td>Class</td><td></td><td>Group ( )</td><td>Group leader’s signature</td><td></td></tr>
<tr><td>Teacher’s signature</td><td></td><td>Date</td><td></td><td></td></tr>
<tr><td colspan="5">Comments:</td></tr>
</table>

3.1.5 Filing the hexagonal side faces

3.1.5.1 Information sheet for filing the hexagonal side faces

<table>
<tr><td>Learning field</td><td colspan="3">Metalworking practical training</td></tr>
<tr><td>Learning scenario 3</td><td>Processing the M10 hexagon nut of transformer low-voltage bushing clamp</td><td>Session teaching time</td><td>8 minutes</td></tr>
<tr><td>Description of the typical working processes</td><td colspan="3">Cutting stock – sawing – filing the end faces – scribing regular hexagon on end faces – filing the hexagonal side faces – drilling a hole – tapping – chamfering</td></tr>
<tr><td>Information collection methods</td><td colspan="3">Teacher's explanation, search on the Internet, or communication among students</td></tr>
<tr><td>Information description</td><td colspan="3">Review the materials provided by the instructor to obtain information about manufacturing the hexagonal side faces, so as to complete the tasks safely and effectively, as indicated in Figure 3.1.5.

Figure 3.1.5 Filing the hexagonal side faces</td></tr>
<tr><td>Requirements for students</td><td colspan="3">Preview the practical training homework sheets before taking the class and use the Internet to find the learning materials related to our practical training. Wear the training uniform throughout the training process. Implement the "4 No-harm Principles" (do not harm yourself, do not harm others, do not get harmed by others, protect others from being harmed) and other safety requirements. Strictly follow the rules and regulations of the training room.</td></tr>
<tr><td>References</td><td colspan="3">Refer to Hexagon Nuts, Style 1 (GB/T 6170—2015).</td></tr>
</table>

3.1.5.2 Planning sheet for filing the hexagonal side faces

<table>
<tr><td>Learning field</td><td colspan="5">Metalworking practical training</td></tr>
<tr><td>Learning scenario 3</td><td colspan="2">Processing the M10 hexagon nut of transformer low-voltage bushing clamp</td><td colspan="2">Session teaching time</td><td>8 minutes</td></tr>
<tr><td>Description of the typical working processes</td><td colspan="5">Cutting stock – sawing – filing the end faces – scribing regular hexagon on end faces – filing the hexagonal side faces – drilling a hole – tapping – chamfering</td></tr>
<tr><td>Planning methods</td><td colspan="5">Group discussion among classmates</td></tr>
<tr><td>Number</td><td colspan="2">Working steps</td><td colspan="3">Notice</td></tr>
<tr><td>1</td><td colspan="2">Preparation before work</td><td colspan="3"></td></tr>
<tr><td>2</td><td colspan="2">Conduct a pre-class meeting, allocate the tasks</td><td colspan="3"></td></tr>
<tr><td>3</td><td colspan="2">Safe operation of filing the hexagonal side faces</td><td colspan="3"></td></tr>
<tr><td>4</td><td colspan="2">Clean up the site</td><td colspan="3"></td></tr>
<tr><td rowspan="3">Plan assessment</td><td>Class</td><td></td><td>Group ( )</td><td>Group leader's signature</td><td></td></tr>
<tr><td>Teacher's signature</td><td></td><td>Date</td><td colspan="2"></td></tr>
<tr><td colspan="5">Comments:</td></tr>
</table>

3.1.5.3　Decision sheet for filing the hexagonal side faces

<table>
<tr><td>Learning field</td><td colspan="5">Metalworking practical training</td></tr>
<tr><td>Learning scenario 3</td><td colspan="2">Processing the M10 hexagon nut of transformer low-voltage bushing clamp</td><td colspan="2">Session teaching time</td><td>8 minutes</td></tr>
<tr><td>Description of the typical working processes</td><td colspan="5">Cutting stock – sawing – filing the end faces – scribing regular hexagon on end faces – filing the hexagonal side faces – drilling a hole – tapping – chamfering</td></tr>
<tr><td colspan="6">Plan comparison</td></tr>
<tr><td>Number</td><td>Feasibility</td><td>Economy</td><td>Operability</td><td>Implementation difficulty</td><td>Integrated evaluation</td></tr>
<tr><td>Plan 1</td><td></td><td></td><td></td><td></td><td></td></tr>
<tr><td>Plan 2</td><td></td><td></td><td></td><td></td><td></td></tr>
<tr><td>Plan 3</td><td></td><td></td><td></td><td></td><td></td></tr>
<tr><td rowspan="3">Decision assessment</td><td>Class</td><td></td><td>Group ( )</td><td>Group leader's signature</td><td></td></tr>
<tr><td>Teacher's signature</td><td></td><td>Date</td><td></td><td></td></tr>
<tr><td colspan="5">Comments:</td></tr>
</table>

3.1.5.4 Implementation sheet for filing the hexagonal side faces

<table>
<tr><td>Learning field</td><td colspan="5">Metalworking practical training</td></tr>
<tr><td>Learning scenario 3</td><td colspan="2">Processing the M10 hexagon nut of transformer low-voltage bushing clamp</td><td>Session teaching time</td><td colspan="2">100 minutes</td></tr>
<tr><td>Description of the typical working processes</td><td colspan="5">Cutting stock – sawing – filing the end faces – scribing regular hexagon on end faces – filing the hexagonal side faces – drilling a hole – tapping – chamfering</td></tr>
<tr><td>Number</td><td>Implementation steps</td><td>Operations and methods</td><td>Process quality requirements</td><td colspan="2">Notice</td></tr>
<tr><td>1</td><td>Preparation before work</td><td></td><td></td><td colspan="2"></td></tr>
<tr><td>2</td><td>Conduct a pre-class meeting, allocate the tasks</td><td></td><td></td><td colspan="2"></td></tr>
<tr><td>3</td><td>Safe operation of filing the hexagonal side faces</td><td>File along the processing boundary line of the hexagon nut end face to form a regular hexagon.</td><td>The required distance between the opposite sides of the regular hexagon is 16mm.</td><td colspan="2">Avoid being harmed by sharp objects.</td></tr>
<tr><td>4</td><td>Clean-up the site</td><td></td><td></td><td colspan="2"></td></tr>
<tr><td>5</td><td></td><td></td><td></td><td colspan="2"></td></tr>
<tr><td colspan="6">Implementation instructions:</td></tr>
<tr><td rowspan="3">Implementation assessment</td><td>Class</td><td></td><td>Group ( )</td><td>Group leader's signature</td><td></td></tr>
<tr><td>Teacher's signature</td><td></td><td>Date</td><td colspan="2"></td></tr>
<tr><td colspan="5">Comments:</td></tr>
</table>

Without the rain, there would never be rainbows. Practice hard, then achieve remarkably.

3.1.5.5 Check sheet for filing the hexagonal side faces

Learning field	Metalworking practical training		
Learning scenario 3	Processing the M10 hexagon nut of transformer low-voltage bushing clamp	Session teaching time	8 minutes
Description of the typical working processes	Cutting stock – sawing – filing the end faces – scribing regular hexagon on end faces – filing the hexagonal side faces – drilling a hole – tapping – chamfering		

Number	Check items	Check criteria	Self-check by students	Check by the teacher
1	Information	Obtain the source of information about manufacturing the hexagonal side faces.		
2	Planning	Formulate relevant plans for manufacturing the hexagonal side faces.		
3	Decision	Compare scores of various plans for manufacturing the hexagonal side faces and select the best solution.		
4	Implementation	Operate each step of manufacturing the hexagonal side faces in accordance with the standards and strictly implement the "4 No-harm Principles".		
5	Check	Check each step carefully, and check whether the theme idea of dedication to work is followed.		

Check assessment	Class		Group ( )	Group leader's signature	
	Teacher's signature		Date		
	Comments:				

3.1.6 Drilling a hole

3.1.6.1 Information sheet for drilling a hole

Learning field	Metalworking practical training		
Learning scenario 3	Processing the M10 hexagon nut of transformer low-voltage bushing clamp	Session teaching time	6 minutes
Description of the typical working processes	Cutting stock – sawing – filing the end faces – scribing regular hexagon on end faces – filing the hexagonal side faces – drilling a hole – tapping – chamfering		
Information collection methods	Teacher's explanation, search on the Internet, or communication among students		
Information description	Review the materials provided by the instructor to obtain information about safe operation of drilling hole, so as to complete the tasks safely and effectively. You can refer to Figure 3.1.6.		
Requirements for students	Preview the practical training homework sheets before taking the class and use the Internet to find the learning materials related to our practical training. Wear the training uniform throughout the training process. Implement the "4 No-harm Principles" (do not harm yourself, do not harm others, do not get harmed by others, protect others from being harmed) and other safety requirements. Strictly follow the rules and regulations of the training room.		
References	Refer to *Hexagon Nuts, Style 1* (GB/T 6170—2015).		

Figure 3.1.6 shows the on-site drilling operation, please point out any violations: ____________________

____

____

Figure 3.1.6 On-site drilling operation

3.1.6.2 Planning sheet for drilling a hole

<table>
<tr><td>Learning field</td><td colspan="5">Metalworking practical training</td></tr>
<tr><td>Learning scenario 3</td><td colspan="3">Processing the M10 hexagon nut of transformer low-voltage bushing clamp</td><td>Session teaching time</td><td>6 minutes</td></tr>
<tr><td>Description of the typical working processes</td><td colspan="5">Cutting stock – sawing – filing the end faces – scribing regular hexagon on end faces – filing the hexagonal side faces – drilling a hole – tapping – chamfering</td></tr>
<tr><td>Planning methods</td><td colspan="5">Group discussion among classmates</td></tr>
<tr><td>Number</td><td colspan="2">Working steps</td><td colspan="3">Notice</td></tr>
<tr><td>1</td><td colspan="2">Preparation before work</td><td colspan="3"></td></tr>
<tr><td>2</td><td colspan="2">Conduct a pre-class meeting, allocate the tasks</td><td colspan="3"></td></tr>
<tr><td>3</td><td colspan="2">Safe operation of drilling a hole</td><td colspan="3">Avoid being harmed by electric shock and the drill bit.</td></tr>
<tr><td>4</td><td colspan="2">Clean up the site</td><td colspan="3"></td></tr>
<tr><td rowspan="3">Plan assessment</td><td>Class</td><td></td><td>Group ( )</td><td>Group leader's signature</td><td></td></tr>
<tr><td>Teacher's signature</td><td></td><td>Date</td><td colspan="2"></td></tr>
<tr><td colspan="5">Comments:</td></tr>
</table>

3.1.6.3 Decision sheet for drilling a hole

<table>
<tr><td colspan="2">Learning field</td><td colspan="5">Metalworking practical training</td></tr>
<tr><td colspan="2">Learning scenario 3</td><td colspan="3">Processing the M10 hexagon nut of transformer low-voltage bushing clamp</td><td>Session teaching time</td><td>6 minutes</td></tr>
<tr><td colspan="2">Description of the typical working processes</td><td colspan="5">Cutting stock – sawing – filing the end faces – scribing regular hexagon on end faces – filing the hexagonal side faces – drilling a hole – tapping – chamfering</td></tr>
<tr><td colspan="7">Plan comparison</td></tr>
<tr><td>Number</td><td>Feasibility</td><td colspan="2">Economy</td><td>Operability</td><td>Implementation difficulty</td><td>Integrated evaluation</td></tr>
<tr><td>Plan 1</td><td></td><td colspan="2"></td><td></td><td></td><td></td></tr>
<tr><td>Plan 2</td><td></td><td colspan="2"></td><td></td><td></td><td></td></tr>
<tr><td>Plan 3</td><td></td><td colspan="2"></td><td></td><td></td><td></td></tr>
<tr><td colspan="2" rowspan="3">Decision assessment</td><td>Class</td><td></td><td>Group ( )</td><td>Group leader's signature</td><td></td></tr>
<tr><td>Teacher's signature</td><td></td><td>Date</td><td></td><td></td></tr>
<tr><td colspan="5">Comments:</td></tr>
</table>

3.1.6.4 Implementation sheet for drilling a hole

Learning field	Metalworking practical training		
Learning scenario 3	Processing the M10 hexagon nut of transformer low-voltage bushing clamp	Session teaching time	20 minutes
Description of the typical working processes	Cutting stock – sawing – filing the end faces – scribing regular hexagon on end faces – filing the hexagonal side faces – drilling a hole – tapping – chamfering		

Number	Implementation steps	Operations and methods	Process quality requirements	Notice
1	Preparation before work			
2	Conduct a pre-class meeting, allocate the tasks			
3	Safe operation of drilling a hole	Punch the punch mark on the center of the circle, drill a hole with a ϕ8.5 drill bit, and ream the hole with a ϕ12.5 drill bit.	The required depth of the punch mark is 1.5mm, hole drilled in the center of the circle. The reaming depth of holes on both end faces should be 1.5mm.	Avoid being harmed by electric shock and the drill bit.
4	Clean up the site			

Implementation instructions:

Implementation assessment	Class		Group ( )	Group leader's signature	
	Teacher's signature		Date		
	Comments:				

3.1.6.5 Check sheet for drilling a hole

<table>
<tr><td colspan="2">Learning field</td><td colspan="6">Metalworking practical training</td></tr>
<tr><td colspan="2">Learning scenario 3</td><td colspan="3">Processing the M10 hexagon nut of transformer low-voltage bushing clamp</td><td colspan="2">Session teaching time</td><td>6 minutes</td></tr>
<tr><td colspan="2">Description of the typical working processes</td><td colspan="6">Cutting stock – sawing – filing the end faces – scribing regular hexagon on end faces – filing the hexagonal side faces – drilling a hole – tapping – chamfering</td></tr>
<tr><td>Number</td><td>Check items</td><td colspan="2">Check criteria</td><td colspan="2">Self-check by students</td><td colspan="2">Check by the teacher</td></tr>
<tr><td>1</td><td>Information</td><td colspan="2">Obtain the source of information about drilling a hole in the M10 hexagon nut.</td><td colspan="2"></td><td colspan="2"></td></tr>
<tr><td>2</td><td>Planning</td><td colspan="2">Formulate relevant plans for drilling a hole.</td><td colspan="2"></td><td colspan="2"></td></tr>
<tr><td>3</td><td>Decision</td><td colspan="2">Compare scores of various hole drilling plans and select the best solution.</td><td colspan="2"></td><td colspan="2"></td></tr>
<tr><td>4</td><td>Implementation</td><td colspan="2">Operate each step of drilling a hole in accordance with the standards and strictly implement the “4 No-harm Principles”.</td><td colspan="2"></td><td colspan="2"></td></tr>
<tr><td>5</td><td>Check</td><td colspan="2">Check each step carefully, and check whether the theme idea of dedication to work is followed.</td><td colspan="2"></td><td colspan="2"></td></tr>
<tr><td colspan="2" rowspan="3">Check assessment</td><td>Class</td><td></td><td>Group ( )</td><td>Group leader’s signature</td><td colspan="2"></td></tr>
<tr><td>Teacher’s signature</td><td></td><td>Date</td><td colspan="3"></td></tr>
<tr><td colspan="6">Comments:</td></tr>
</table>

3.1.7 Tapping

3.1.7.1 Information sheet for tapping

Learning field	Metalworking practical training		
Learning scenario 3	Processing the M10 hexagon nut of transformer low-voltage bushing clamp	Session teaching time	6 minutes
Description of the typical working processes	Cutting stock – sawing – filing the end faces – scribing regular hexagon on end faces – filing the hexagonal side faces – drilling a hole – tapping – chamfering		
Information collection methods	Teacher's explanation, search on the Internet, or communication among students		
Information description	Review the materials provided by the instructor to obtain information about manufacturing the tapping of the hexagon nut, so as to complete the tasks safely and effectively.		
Requirements for students	Preview the practical training homework sheets before taking the class and use the Internet to find the learning materials related to our practical training. Wear the training uniform throughout the training process. Implement the "4 No-harm Principles" (do not harm yourself, do not harm others, do not get harmed by others, protect others from being harmed) and other safety requirements. Strictly follow the rules and regulations of the training room.		
References	Refer to *Hexagon Nuts, Style 1* (GB/T 6170—2015) and *Technical Code for Safety of Temporary Electrification on Construction Site* (JGJ 46—2005) Operation of tapping		

Technical key points

Which of the following statements is/are correct ( ).

A.Keep the tap perpendicular to the end face of the nut when tapping, with the center of the tap overlapping with the center of the hole.

B.Hold tightly two ends of the tap wrench with both hands, and rotate the tap wrench clockwise, reversing 1~2 turns for every 4~5 turns.

C.First, use the first tap, then the second tap for tapping.

D.When the tap is not perpendicular to the hole surface, tap the tap with a hammer.

E.If you feel a lot of resistance when tapping, you can forcefully twist the tap to keep it moving.

F.The speed of filing is generally 30~60 times per minute.

G.The speed of sawing is generally 60 times per minute.

H.Do not operate the drill when wearing gloves.

I. The electricity to the bench drill should be cut off when replacing the drill bit.

3.1.7.2 Planning sheet for tapping

<table>
<tr><td>Learning field</td><td colspan="5">Metalworking practical training</td></tr>
<tr><td>Learning scenario 3</td><td colspan="3">Processing the M10 hexagon nut of transformer low-voltage bushing clamp</td><td>Session teaching time</td><td>6 minutes</td></tr>
<tr><td>Description of the typical working processes</td><td colspan="5">Cutting stock – sawing – filing the end faces – scribing regular hexagon on end faces – filing the hexagonal side faces – drilling a hole – tapping – chamfering</td></tr>
<tr><td>Planning methods</td><td colspan="5">Group discussion among classmates</td></tr>
<tr><td>Number</td><td>Working steps</td><td colspan="4">Notice</td></tr>
<tr><td>1</td><td>Preparation before work</td><td colspan="4"></td></tr>
<tr><td>2</td><td>Conduct a pre-class meeting, allocate the tasks</td><td colspan="4"></td></tr>
<tr><td>3</td><td>Safe operation of tapping</td><td colspan="4"></td></tr>
<tr><td>4</td><td>Clean up the site</td><td colspan="4"></td></tr>
<tr><td rowspan="3">Plan assessment</td><td>Class</td><td></td><td>Group ( )</td><td>Group leader's signature</td><td></td></tr>
<tr><td>Teacher's signature</td><td></td><td>Date</td><td colspan="2"></td></tr>
<tr><td colspan="5">Comments:</td></tr>
</table>

3.1.7.3 Decision sheet for tapping

<table>
<tr><td>Learning field</td><td colspan="5">Metalworking practical training</td></tr>
<tr><td>Learning scenario 3</td><td colspan="3">Processing the M10 hexagon nut of transformer low-voltage bushing clamp</td><td>Session teaching time</td><td>6 minutes</td></tr>
<tr><td>Description of the typical working processes</td><td colspan="5">Cutting stock – sawing – filing the end faces – scribing regular hexagon on end faces – filing the hexagonal side faces – drilling a hole – tapping – chamfering</td></tr>
<tr><td colspan="6">Plan comparison</td></tr>
<tr><td>Number</td><td>Feasibility</td><td>Economy</td><td>Operability</td><td>Implementation difficulty</td><td>Integrated evaluation</td></tr>
<tr><td>Plan 1</td><td></td><td></td><td></td><td></td><td></td></tr>
<tr><td>Plan 2</td><td></td><td></td><td></td><td></td><td></td></tr>
<tr><td>Plan 3</td><td></td><td></td><td></td><td></td><td></td></tr>
<tr><td rowspan="3">Decision assessment</td><td>Class</td><td></td><td>Group ( )</td><td>Group leader's signature</td><td></td></tr>
<tr><td>Teacher's signature</td><td></td><td>Date</td><td colspan="2"></td></tr>
<tr><td colspan="5">Comments:</td></tr>
</table>

3.1.7.4 Implementation sheet for tapping

<table>
<tr><td>Learning field</td><td colspan="4">Metalworking practical training</td></tr>
<tr><td>Learning scenario 3</td><td colspan="2">Processing the M10 hexagon nut of transformer low-voltage bushing clamp</td><td>Session teaching time</td><td>20 minutes</td></tr>
<tr><td>Description of the typical working processes</td><td colspan="4">Cutting stock – sawing – filing the end faces – scribing regular hexagon on end faces – filing the hexagonal side faces – drilling a hole – tapping – chamfering</td></tr>
<tr><td>Number</td><td>Implementation steps</td><td>Operations and methods</td><td>Process quality requirements</td><td>Notice</td></tr>
<tr><td>1</td><td>Preparation before work</td><td></td><td></td><td></td></tr>
<tr><td>2</td><td>Conduct a pre-class meeting, allocate the tasks</td><td></td><td></td><td></td></tr>
<tr><td>3</td><td>Safe operation of tapping</td><td>Use M10 taps (first tap, second tap) for tapping. During the tapping process, the force of both hands should be balanced, and the rotation should be balanced. Reverse the tap by 1/4 turn for each 1/2 ~ 1 turn to remove the swarf, preventing the swarf from blocking the flute, which can cause damage or breakage of the tap.</td><td>Smooth screw thread.</td><td>Avoid being harmed by sharp objects.</td></tr>
<tr><td>4</td><td>Clean up the site</td><td></td><td></td><td></td></tr>
<tr><td colspan="5">Implementation instructions:</td></tr>
</table>

<table>
<tr><td rowspan="3">Implementation assessment</td><td>Class</td><td></td><td>Group ( )</td><td>Group leader's signature</td><td></td></tr>
<tr><td>Teacher's signature</td><td></td><td>Date</td><td colspan="2"></td></tr>
<tr><td colspan="5">Comments:</td></tr>
</table>

3.1.7.5 Check sheet for tapping

Learning field	Metalworking practical training		
Learning scenario 3	Processing the M10 hexagon nut of transformer low-voltage bushing clamp	Session teaching time	6 minutes
Description of the typical working processes	Cutting stock – sawing – filing the end faces – scribing regular hexagon on end faces – filing the hexagonal side faces – drilling a hole – tapping – chamfering		

Number	Check items	Check criteria	Self-check by students	Check by the teacher
1	Information	Obtain the source of information about the operation of tapping of the M10 hexagon nut.		
2	Planning	Formulate relevant plans for tapping.		
3	Decision	Compare scores of various tapping plans and select the best solution.		
4	Implementation	Operate each step of tapping in accordance with the standards and strictly implement the "4 No-harm Principles".		
5	Check	Check each step carefully, and check whether the theme idea of dedication to work is followed.		

Check assessment	Class		Group ( )	Group leader's signature	
	Teacher's signature		Date		
	Comments:				

3.1.8 Chamfering

3.1.8.1 Information sheet for chamfering

Learning field	Metalworking practical training		
Learning scenario 3	Processing the M10 hexagon nut of transformer low-voltage bushing clamp	Session teaching time	6 minutes
Description of the typical working processes	Cutting stock – sawing – filing the end faces – scribing regular hexagon on end faces – filing the hexagonal side faces – drilling a hole – tapping – chamfering		
Information collection methods	Teacher's explanation, search on the Internet, or communication among students		
Information description	Review the materials provided by the instructor to obtain information about operating the chamfer of the hexagon nut, so as to complete the tasks safely and effectively.		
Requirements for students	Preview the practical training homework sheets before taking the class and use the Internet to find the learning materials related to our practical training. Wear the training uniform throughout the training process. Implement the "4 No-harm Principles" (do not harm yourself, do not harm others, do not get harmed by others, protect others from being harmed) and other safety requirements. Strictly follow the rules and regulations of the training room.		
References	Refer to *Hexagon Nuts, Style 1* (GB/T 6170—2015).		

Technical key points

1. Which of the following tools cannot be tapped by a hammer (   )?

A. Scribing compass   B. Anvil   C. Scriber   D. Vernier caliper

2. Which of the following statements is/are correct (   )?

A.The tighter the saw blade is installed on the bow saw, the better it is.

B.When the sawing workpiece is clamped by a vise, the sawing position should not be overextended.

C.When the sawing of the workpiece has started or will soon be cut off, it must be gently pushed and cut to prevent it from slipping out and causing hand injury or breaking the saw blade.

D.The sawing workpiece should be tightly clamped, and the sawing of the workpiece should be smooth.

E.The file should be installed with the handle.

F.The file cannot be placed outside the working table.

G.Prying, smashing, and tapping other items with the file is not allowed.

3.1.8.2 Planning sheet for chamfering

<table>
<tr><td>Learning field</td><td colspan="5">Metalworking practical training</td></tr>
<tr><td>Learning scenario 3</td><td colspan="3">Processing the M10 hexagon nut of transformer low-voltage bushing clamp</td><td>Session teaching time</td><td>6 minutes</td></tr>
<tr><td>Description of the typical working processes</td><td colspan="5">Cutting stock – sawing – filing the end faces – scribing regular hexagon on end faces – filing the hexagonal side faces – drilling a hole – tapping – chamfering</td></tr>
<tr><td>Planning methods</td><td colspan="5">Group discussion among classmates</td></tr>
<tr><td>Number</td><td colspan="2">Working steps</td><td colspan="3">Notice</td></tr>
<tr><td>1</td><td colspan="2">Preparation before work</td><td colspan="3"></td></tr>
<tr><td>2</td><td colspan="2">Conduct a pre-class meeting, allocate the tasks</td><td colspan="3"></td></tr>
<tr><td>3</td><td colspan="2">Safe operation of tapping</td><td colspan="3"></td></tr>
<tr><td>4</td><td colspan="2">Clean up the site</td><td colspan="3"></td></tr>
<tr><td rowspan="3">Plan assessment</td><td>Class</td><td></td><td>Group ( )</td><td>Group leader's signature</td><td></td></tr>
<tr><td>Teacher's signature</td><td></td><td>Date</td><td colspan="2"></td></tr>
<tr><td colspan="5">Comments:</td></tr>
</table>

3.1.8.3 Decision sheet for chamfering

<table>
<tr><td>Learning field</td><td colspan="5">Metalworking practical training</td></tr>
<tr><td>Learning scenario 3</td><td colspan="3">Processing the M10 hexagon nut of transformer low-voltage bushing clamp</td><td>Session teaching time</td><td>6 minutes</td></tr>
<tr><td>Description of the typical working processes</td><td colspan="5">Cutting stock – sawing – filing the end faces – scribing regular hexagon on end faces – filing the hexagonal side faces – drilling a hole – tapping – chamfering</td></tr>
<tr><td colspan="6">Plan comparison</td></tr>
<tr><td>Number</td><td>Feasibility</td><td>Economy</td><td>Operability</td><td>Implementation difficulty</td><td>Integrated evaluation</td></tr>
<tr><td>Plan 1</td><td></td><td></td><td></td><td></td><td></td></tr>
<tr><td>Plan 2</td><td></td><td></td><td></td><td></td><td></td></tr>
<tr><td>Plan 3</td><td></td><td></td><td></td><td></td><td></td></tr>
<tr><td rowspan="3">Decision assessment</td><td>Class</td><td></td><td>Group ( )</td><td>Group leader's signature</td><td></td></tr>
<tr><td>Teacher's signature</td><td></td><td>Date</td><td colspan="2"></td></tr>
<tr><td colspan="5">Comments:</td></tr>
</table>

3.1.8.4 Implementation sheet for chamfering

<table>
<tr><td colspan="2">Learning field</td><td colspan="3">Metalworking practical training</td></tr>
<tr><td colspan="2">Learning scenario 3</td><td colspan="2">Processing the M10 hexagon nut of transformer low-voltage bushing clamp</td><td>Session teaching time</td><td>30 minutes</td></tr>
<tr><td colspan="2">Description of the typical working processes</td><td colspan="4">Cutting stock – sawing – filing the end faces – scribing regular hexagon on end faces – filing the hexagonal side faces – drilling a hole – tapping – chamfering</td></tr>
<tr><td>Number</td><td>Implementation steps</td><td>Operations and methods</td><td colspan="2">Process quality requirements</td><td>Notice</td></tr>
<tr><td>1</td><td>Preparation before work</td><td></td><td colspan="2"></td><td></td></tr>
<tr><td>2</td><td>Conduct a pre-class meeting, allocate the tasks</td><td></td><td colspan="2"></td><td></td></tr>
<tr><td>3</td><td>Safe operation of chamfering</td><td>File off the six corners of the two end faces.</td><td colspan="2">It is required that each corner being filed off should form an equilateral triangle.</td><td>Avoid being harmed by sharp objects.</td></tr>
<tr><td>4</td><td>Clean up the site</td><td>1.After the site work, the person in charge should turn off the electricity and check the tools.
2.Clean up the site.</td><td colspan="2">Conduct safety check, check and collect the tools, and clean up the site as required.</td><td>Avoid being harmed by iron filings and sharp objects.</td></tr>
<tr><td colspan="6">Implementation instructions:</td></tr>
<tr><td rowspan="3">Implementation assessment</td><td>Class</td><td></td><td>Group ( )</td><td>Group leader's signature</td><td></td></tr>
<tr><td>Teacher's signature</td><td></td><td>Date</td><td colspan="2"></td></tr>
<tr><td colspan="5">Comments:</td></tr>
</table>

3.1.8.5 Check sheet for chamfering

Learning field	Metalworking practical training		
Learning scenario 3	Processing the M10 hexagon nut of transformer low-voltage bushing clamp	Session teaching time	6 minutes
Description of the typical working processes	Cutting stock – sawing – filing the end faces – scribing regular hexagon on end faces – filing the hexagonal side faces – drilling a hole – tapping – chamfering		

Number	Check items	Check criteria	Self-check by students	Check by the teacher
1	Information	Obtain the source of information about the operation of chamfering of the M10 hexagon nut.		
2	Planning	Formulate relevant plans for chamfering.		
3	Decision	Compare scores of various chamfering operation plans and select the best solution.		
4	Implementation	Operate each step of chamfering in accordance with the standards and strictly implement the “4 No-harm Principles”.		
5	Check	Check each step carefully, and check whether the theme idea of dedication to work is followed.		

<table>
<tr><td rowspan="3">Check assessment</td><td>Class</td><td></td><td>Group ( )</td><td>Group leader’s signature</td><td></td></tr>
<tr><td>Teacher’s signature</td><td></td><td>Date</td><td colspan="2"></td></tr>
<tr><td colspan="5">Comments:</td></tr>
</table>

3.2 List of materials and tools for processing the M10 hexagon nut of transformer low-voltage bushing clamp learning scenario

Learning field	Metalworking practical training				
Learning scenario 3	Processing the M10 hexagon nut of transformer low-voltage bushing clamp		Session teaching time	540 minutes	
Description of the typical working processes	Cutting stock – sawing – filing the end faces – scribing regular hexagon on end faces – filing the hexagonal side faces – drilling a hole – tapping – chamfering				
Number	Name	Function	Quantity	Model	User
Class		Group ( )	Group leader's signature		
Teacher's signature		Date			

3.3 Teaching feedback sheet for processing the M10 hexagon nut of transformer low-voltage bushing clamp learning scenario

<table>
<tr><td>Learning field</td><td colspan="5">Metalworking practical training</td></tr>
<tr><td>Learning scenario 3</td><td colspan="2">Processing the M10 hexa-gon nut of transformer low-voltage bushing clamp</td><td colspan="2">Session teaching time</td><td>540 minutes</td></tr>
<tr><td>Description of the typical working processes</td><td colspan="5">Cutting stock – sawing – filing the end faces – scribing regular hexagon on end faces – filing the hexagonal side faces – drilling a hole – tapping – chamfering</td></tr>
<tr><td rowspan="6">Survey items</td><td>Number</td><td colspan="3">Survey contents</td><td>Description of the reasons</td></tr>
<tr><td></td><td colspan="3"></td><td></td></tr>
<tr><td></td><td colspan="3"></td><td></td></tr>
<tr><td></td><td colspan="3"></td><td></td></tr>
<tr><td></td><td colspan="3"></td><td></td></tr>
<tr><td></td><td colspan="3"></td><td></td></tr>
<tr><td colspan="6">Your feedback on how to improve the teaching of this course:</td></tr>
<tr><td>Survey information</td><td colspan="2">Respondent's name</td><td></td><td>Survey date</td><td></td></tr>
</table>

3.4 Grouping sheet for processing the M10 hexagon nut of transformer low-voltage bushing clamp learning scenario

Learning field	Metalworking practical training					
Learning scenario 3	Processing the M10 hexagon nut of transformer low-voltage bushing clamp			Session teaching time	540 minutes	
Description of the typical working processes	Cutting stock – sawing – filing the end faces – scribing regular hexagon on end faces – filing the hexagonal side faces – drilling a hole – tapping – chamfering					
Grouping	Group	Group leader	Group members			
Grouping instructions						
Class		Teacher's signature		Date		

3.5 Assessment sheet for processing the M10 hexagon nut of transformer low-voltage bushing clamp learning scenario

Learning field	Metalworking practical training			
Learning scenario 3	Processing the M10 hexagon nut of transformer low-voltage bushing clamp	Session teaching time		20 minutes
Description of the typical working processes	Cutting stock – sawing – filing the end faces – scribing regular hexagon on end faces – filing the hexagonal side faces – drilling a hole – tapping – chamfering			
Assessment items	Assessment subitems	Self-assessment by students 20%	Intra-group assessment 30%	Assessment by the teacher 50%
Cutting stock	1. Information: 10 points; 2. Planning: 10 points; 3. Decision: 10 points; 4. Implementation: 40 points; 5. Check: 10 points; 6. Work attitude: 20 points.			
Sawing	1. Information: 10 points; 2. Planning: 10 points; 3. Decision: 10 points; 4. Implementation: 40 points; 5. Check: 10 points; 6. Work attitude: 20 points.			
Filing the end faces	1. Information: 10 points; 2. Planning: 10 points; 3. Decision: 10 points; 4. Implementation: 40 points; 5. Check: 10 points; 6. Work attitude: 20 points.			
Scribing regular hexagon on end faces	1. Information: 10 points; 2. Planning: 10 points; 3. Decision: 10 points; 4. Implementation: 40 points; 5. Check: 10 points; 6. Work attitude: 20 points.			
Filing the hexagonal side faces	1. Information: 10 points; 2. Planning: 10 points; 3. Decision: 10 points; 4. Implementation: 40 points; 5. Check: 10 points; 6. Work attitude: 20 points.			

Continue

<table>
<tr><td>Assessment items</td><td colspan="2">Assessment subitems</td><td colspan="2">Self-assessment by students 20%</td><td>Intra-group assessment 30%</td><td>Assessment by the teacher 50%</td></tr>
<tr><td>Drilling a hole</td><td colspan="2">1. Information: 10 points;
2. Planning: 10 points;
3. Decision: 10 points;
4. Implementation: 40 points;
5. Check: 10 points;
6. Work attitude: 20 points.</td><td colspan="2"></td><td></td><td></td></tr>
<tr><td>Tapping</td><td colspan="2">1. Information: 10 points;
2. Planning: 10 points;
3. Decision: 10 points;
4. Implementation: 40 points;
5. Check: 10 points;
6. Work attitude: 20 points.</td><td colspan="2"></td><td></td><td></td></tr>
<tr><td>Chamfering</td><td colspan="2">1. Information: 10 points;
2. Planning: 10 points;
3. Decision: 10 points;
4. Implementation: 40 points;
5. Check: 10 points;
6. Work attitude: 20 points.</td><td colspan="2"></td><td></td><td></td></tr>
<tr><td>Final results</td><td colspan="2">1. Information: 10 points;
2. Planning: 10 points;
3. Decision: 10 points;
4. Implementation: 40 points;
5. Check: 10 points;
6. Work attitude: 20 points.</td><td colspan="2"></td><td></td><td></td></tr>
<tr><td rowspan="3">Assessment</td><td>Class</td><td colspan="2"></td><td>Group ( )</td><td>Group leader's signature</td><td></td></tr>
<tr><td>Teacher's signature</td><td colspan="2"></td><td>Date</td><td colspan="2"></td></tr>
<tr><td colspan="6">Comments:</td></tr>
</table>

3.6 Teacher's implementation planning sheet

Learning field	Metalworking practical training					
Learning scenario 3	Processing the M10 hexagon nut of transformer low-voltage bushing clamp		Session teaching time		540 minutes	
Description of the typical working processes	Cutting stock – sawing – filing the end faces – scribing regular hexagon on end faces – filing the hexagonal side faces – drilling a hole – tapping – chamfering					
Number	Working and learning steps	Teaching time	Tools	Locations	Methods	Notes
Class		Teacher's signature			Date	

学习情境 4

电气设备接地引下线的焊接

4.0 学习性工作任务单

<table>
<tr><td>学习场</td><td colspan="6">金　工　实　训</td></tr>
<tr><td>学习情境 4</td><td colspan="4">电气设备接地引下线的焊接</td><td>总学时</td><td>660 分钟</td></tr>
<tr><td>典型工作过程描述</td><td colspan="6">焊接引弧技能训练—焊接运条技能训练—焊接收尾技能训练—下料—接地引下线搭接平焊</td></tr>
<tr><td>学习目标</td><td colspan="6">1. 能正确使用焊接工具。
2. 能安全规范操作电焊机。
3. 掌握焊接的引弧、运条和收尾的安全操作技能。
4. 掌握平焊的基本技能。</td></tr>
<tr><td>任务描述</td><td colspan="6">把两块尺寸为 4mm×40mm×200mm 的扁铁搭接在一起，经过手工电弧焊接端面 A、端面 B 和端面 C，如图 4.0.1 所示，使其达到电气设备接地连接的要求。
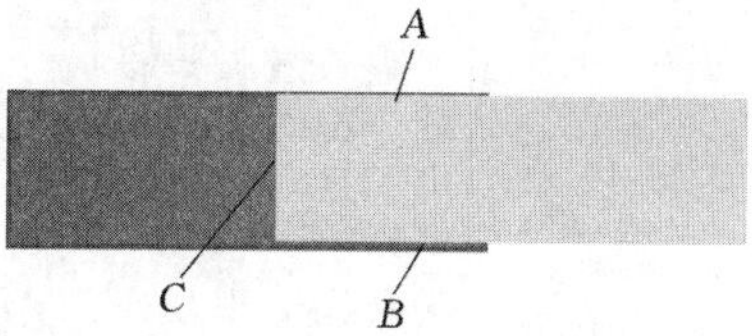

图 4.0.1　三面焊接</td></tr>
<tr><td rowspan="2">学时安排</td><td>资讯</td><td>计划</td><td>决策</td><td>实施</td><td>检查</td><td>评价</td></tr>
<tr><td>30 分钟</td><td>30 分钟</td><td>30 分钟</td><td>520 分钟</td><td>30 分钟</td><td>20 分钟</td></tr>
<tr><td>对学生的要求</td><td colspan="6">课前充分预习实训作业单，实训过程中穿着实训服装，贯彻落实自己不伤害自己、自己不伤害他人、自己不被他人伤害、保护他人不被伤害的“四不伤害”安全要求，注意严格遵守实训室的各项规章制度。</td></tr>
<tr><td>参考资料</td><td colspan="6">参考《电气装置安装工程　接地装置施工及验收规范》（GB 50169—2016）和《电弧焊机通用技术条件》（GB/T 8118—2010）。
焊接三步骤　焊接的操作</td></tr>
</table>

不经历风雨，怎么见彩虹。苦练千百次，操作如有神。

4.1 教学引导文

学习场	金工实训	学习情境4	电气设备接地引下线的焊接	参照系	《电气装置安装工程 接地装置施工及验收规范》(GB 50169—2016)	
普适性工作过程 典型工作过程	资讯	计划	决策	实施	检查	评价
焊接引弧技能训练	查看《施工现场临时用电安全技术规范》(JGJ 46—2005),获取施工现场用电规范的有关信息,便于安全有效地完成任务。	通过同学之间小组讨论,拟订用电安全规范计划,完成4.1.1.2焊接引弧技能训练计划单。	计划的可行性、经济性、可操作性和实施难度几方面综合分析,选取最佳方案。	查找有关资料,完成4.1.1.4焊接引弧技能训练实施单。	按照4.1.1.5焊接引弧技能训练检查单的检查项目和标准,进行学生自查和教师检查。	根据评价项目配置分值,进行自评、互评和教师评价,完成4.5电气设备接地引下线的焊接学习情境评价单。
焊接运条技能训练	查看《弧焊设备 第1部分:焊接电源》(GB/T 15579.1—2013),获取施工现场点焊接用电技术要求的有关信息,便于安全有效地完成任务。	通过同学之间小组讨论,拟订用电安全规范计划,完成4.1.2.2焊接运条技能训练计划单。	计划的可行性、经济性、可操作性和实施难度几方面综合分析,选取最佳方案。	查找有关资料,完成4.1.2.4焊接运条技能训练实施单。	按照4.1.2.5焊接运条技能训练检查单的检查项目和标准,进行学生自查和教师检查。	根据评价项目配置分值,进行自评、互评和教师评价,完成4.5电气设备接地引下线的焊接学习情境评价单。
焊接收尾技能训练	查看《施工现场临时用电安全技术规范》(JGJ 46—2005),获取施工现场用电规范的有关信息,便于安全有效地完成任务。	通过同学之间小组讨论,拟订用电安全规范计划,完成4.1.3.2焊接收尾技能训练计划单。	计划的可行性、经济性、可操作性和实施难度几方面综合分析,选取最佳方案。	查找有关资料,完成4.1.3.4焊接收尾技能训练实施单。	按照4.1.3.5焊接收尾技能训练检查单的检查项目和标准,进行学生自查和教师检查。	根据评价项目配置分值,进行自评、互评和教师评价,完成4.5电气设备接地引下线的焊接学习情境评价单。

续表

学习场	金工实训	学习情境4	电气设备接地引下线的焊接	参照系	《电气装置安装工程　接地装置施工及验收规范》（GB 50169—2016）	
普适性工作过程 典型工作过程	资讯	计划	决策	实施	检查	评价
下料	查看《施工现场临时用电安全技术规范》（JGJ 46—2005），获取施工现场用电规范的有关信息，便于安全有效地完成任务。	通过同学之间小组讨论，拟订用电安全规范计划，完成4.1.4.2下料计划单。	计划的可行性、经济性、可操作性和实施难度几方面综合分析，选取最佳方案。	查找有关资料，完成4.1.4.4下料实施单。	按照4.1.4.5下料检查单的检查项目和标准，进行学生自查和教师检查。	根据评价项目配置分值，进行自评、互评和教师评价，完成4.5电气设备接地引下线的焊接学习情境评价单。
接地引下线搭接平焊	查看《弧焊设备　第1部分：焊接电源》（GB/T 15579.1—2013），获取施工现场点焊接用电技术要求的有关信息，便于安全有效地完成任务，参考图4.1.1。	通过同学之间小组讨论，拟订用电安全规范计划，完成4.1.5.2接地引下线搭接平焊计划单。	计划的可行性、经济性、可操作性和实施难度几方面综合分析，选取最佳方案。	查找有关资料，完成4.1.5.4接地引下线搭接平焊实施单。	按照4.1.5.5接地引下线搭接平焊检查单的检查项目和标准，进行学生自查和教师检查。	根据评价项目配置分值，进行自评、互评和教师评价，完成4.5电气设备接地引下线的焊接学习情境评价单。

图4.1.1　现场作业指导书

学习情境4-3

不经历风雨，怎么见彩虹。苦练千百次，操作如有神。

4.1.1　焊接引弧技能训练

4.1.1.1　焊接引弧技能训练资讯单

学习场	金　工　实　训		
学习情境 4	电气设备接地引下线的焊接	本环节学时	6 分钟
典型工作过程描述	焊接引弧技能训练—焊接运条技能训练—焊接收尾技能训练—下料—接地引下线搭接平焊		
搜集资讯的方式	教师的讲解、互联网查询或者同学间交流		
资讯描述	查看教师提供的资料，获取手工电弧焊接引弧安全操作的技能要求和有关信息，便于安全有效地完成任务，参考图 4.1.2。  图 4.1.2　引弧训练		
对学生的要求	课前充分预习实训作业单，实训过程中穿着实训服装，贯彻落实自己不伤害自己、自己不伤害他人、自己不被他人伤害、保护他人不被伤害的“四不伤害”安全要求，注意严格遵守实训室的各项规章制度。		
参考资料	参考《施工现场临时用电安全技术规范》(JGJ 46—2005) 和《电弧焊焊接工艺规程》(GB/T 19867.1—2005)。		

4.1.1.2 焊接引弧技能训练计划单

<table>
<tr><td>学习场</td><td colspan="5">金 工 实 训</td></tr>
<tr><td>学习情境 4</td><td colspan="3">电气设备接地引下线的焊接</td><td>本环节学时</td><td>6 分钟</td></tr>
<tr><td>典型工作过程描述</td><td colspan="5">焊接引弧技能训练—焊接运条技能训练—焊接收尾技能训练—下料—接地引下线搭接平焊</td></tr>
<tr><td>计划制订的方式</td><td colspan="5">同学之间小组讨论</td></tr>
<tr><td>序号</td><td>工 作 步 骤</td><td colspan="4">注 意 事 项</td></tr>
<tr><td>1</td><td>作业前准备</td><td colspan="4"></td></tr>
<tr><td>2</td><td>开班前会，布置任务</td><td colspan="4"></td></tr>
<tr><td>3</td><td>安全操作引弧技能训练</td><td colspan="4"></td></tr>
<tr><td>4</td><td>清理现场</td><td colspan="4"></td></tr>
<tr><td rowspan="3">计划评价</td><td>班级</td><td></td><td>第（ ）组</td><td>组长签字</td><td></td></tr>
<tr><td>教师签字</td><td></td><td>日期</td><td colspan="2"></td></tr>
<tr><td colspan="5">评语：</td></tr>
</table>

4.1.1.3 焊接引弧技能训练决策单

<table>
<tr><td colspan="2">学习场</td><td colspan="5">金 工 实 训</td></tr>
<tr><td colspan="2">学习情境 4</td><td colspan="3">电气设备接地引下线的焊接</td><td>本环节学时</td><td>6 分钟</td></tr>
<tr><td colspan="2">典型工作过程描述</td><td colspan="5">焊接引弧技能训练—焊接运条技能训练—焊接收尾技能训练—下料—接地引下线搭接平焊</td></tr>
<tr><td colspan="7">计 划 对 比</td></tr>
<tr><td>序号</td><td colspan="2">计划的可行性</td><td>计划的经济性</td><td>计划的可操作性</td><td>计划的实施难度</td><td>综合评价</td></tr>
<tr><td>方案 1</td><td colspan="2"></td><td></td><td></td><td></td><td></td></tr>
<tr><td>方案 2</td><td colspan="2"></td><td></td><td></td><td></td><td></td></tr>
<tr><td>方案 3</td><td colspan="2"></td><td></td><td></td><td></td><td></td></tr>
<tr><td colspan="2" rowspan="3">决策评价</td><td>班级</td><td></td><td>第（ ）组</td><td>组长签字</td><td></td></tr>
<tr><td>教师签字</td><td></td><td>日期</td><td colspan="2"></td></tr>
<tr><td colspan="5">评语：</td></tr>
</table>

学习情境 4-6

4.1.1.4 焊接引弧技能训练实施单

<table>
<tr><td>学习场</td><td colspan="4">金　工　实　训</td></tr>
<tr><td>学习情境 4</td><td colspan="2">电气设备接地引下线的焊接</td><td>本环节学时</td><td>60 分钟</td></tr>
<tr><td>典型工作过程描述</td><td colspan="4">焊接引弧技能训练—焊接运条技能训练—焊接收尾技能训练—下料—接地引下线搭接平焊</td></tr>
<tr><td>序号</td><td>实施步骤</td><td>工作内容及方法</td><td>工艺质量要求</td><td>注意事项</td></tr>
<tr><td>1</td><td>工作前准备</td><td></td><td></td><td></td></tr>
<tr><td>2</td><td>开班前会，布置任务</td><td></td><td></td><td></td></tr>
<tr><td>3</td><td>安全操作引弧技能训练</td><td>1. 敲击法：使焊条与焊件表面垂直地接触，当焊条的末端与焊件的表面轻轻一碰，便迅速提起焊条并保持一定的距离，立即引燃电弧。
2. 划擦法：先将焊条末端对准焊件，然后将焊条在焊件表面划擦一下，当电弧引燃后趁金属还没有开始大量熔化的一瞬间，立即使焊条末端与被焊表面的距离维持在 2～4mm，电弧就能稳定地燃烧。</td><td>操作时必须掌握好手腕上下动作的时间和距离，保证电弧能稳定地燃烧。</td><td>人身触电、电弧伤害、毒气和烟尘伤害、高温伤害</td></tr>
<tr><td>4</td><td>清理现场</td><td></td><td></td><td></td></tr>
<tr><td colspan="5">实施说明：</td></tr>
<tr><td rowspan="3">实施评价</td><td>班级</td><td></td><td>第（　　）组</td><td>组长签字</td></tr>
<tr><td>教师签字</td><td></td><td>日期</td><td></td></tr>
<tr><td colspan="4">评语：</td></tr>
</table>

4.1.1.5　焊接引弧技能训练检查单

<table>
<tr><td colspan="2">学习场</td><td colspan="6">金　工　实　训</td></tr>
<tr><td colspan="2">学习情境 4</td><td colspan="3">电气设备接地引下线的焊接</td><td colspan="2">本环节学时</td><td>6 分钟</td></tr>
<tr><td colspan="2">典型工作过程描述</td><td colspan="6">焊接引弧技能训练—焊接运条技能训练—焊接收尾技能训练—下料—接地引下线搭接平焊</td></tr>
<tr><td>序号</td><td>检查项目</td><td colspan="3">检 查 标 准</td><td colspan="2">学生自查</td><td>教师检查</td></tr>
<tr><td>1</td><td>资讯环节</td><td colspan="3">获得电弧焊接引弧安全操作技能的有关信息的来源。</td><td colspan="2"></td><td></td></tr>
<tr><td>2</td><td>计划环节</td><td colspan="3">制定电弧焊接引弧安全操作技能步骤的各个方案。</td><td colspan="2"></td><td></td></tr>
<tr><td>3</td><td>决策环节</td><td colspan="3">对各种电弧焊接引弧安全操作技能的方案衡量对比评分，获取最佳方案。</td><td colspan="2"></td><td></td></tr>
<tr><td>4</td><td>实施环节</td><td colspan="3">电弧焊接引弧安全操作技能训练过程中各个环节按照标准执行，贯彻落实“四不伤害”。</td><td colspan="2"></td><td></td></tr>
<tr><td>5</td><td>检查环节</td><td colspan="3">各个环节逐一检查，是否有爱岗敬业的主题思想。</td><td colspan="2"></td><td></td></tr>
<tr><td colspan="2" rowspan="3">检查评价</td><td>班级</td><td></td><td>第（　　）组</td><td>组长签字</td><td colspan="2"></td></tr>
<tr><td>教师签字</td><td></td><td>日期</td><td colspan="3"></td></tr>
<tr><td colspan="6">评语：</td></tr>
</table>

4.1.2　焊接运条技能训练

4.1.2.1　焊接运条技能训练资讯单

学习场	金　工　实　训		
学习情境 4	电气设备接地引下线的焊接	本环节学时	6 分钟
典型工作过程描述	焊接引弧技能训练—焊接运条技能训练—焊接收尾技能训练—下料—接地引下线搭接平焊		
搜集资讯的方式	教师的讲解、互联网查询或者同学间交流		
资讯描述	查看教师提供的资料，获取手工电弧焊接运条安全操作的技能要求和有关信息，便于安全有效地完成任务，参考图 4.1.3。  图 4.1.3　运条技能训练		
对学生的要求	课前充分预习实训作业单，实训过程中穿着实训服装，贯彻落实自己不伤害自己、自己不伤害他人、自己不被他人伤害、保护他人不被伤害的“四不伤害”安全要求注意事项，严格遵守实训室的各项规章制度。		
参考资料	参考《施工现场临时用电安全技术规范》（JGJ 46—2005）和《电弧焊焊接工艺规程》（GB/T 19867.1—2005）。		

4.1.2.2 焊接运条技能训练计划单

<table>
<tr><td colspan="2">学习场</td><td colspan="5">金 工 实 训</td></tr>
<tr><td colspan="2">学习情境 4</td><td colspan="3">电气设备接地引下线的焊接</td><td>本环节学时</td><td>6 分钟</td></tr>
<tr><td colspan="2">典型工作过程描述</td><td colspan="5">焊接引弧技能训练—焊接运条技能训练—焊接收尾技能训练—下料—接地引下线搭接平焊</td></tr>
<tr><td colspan="2">计划制订的方式</td><td colspan="5">同学之间小组讨论</td></tr>
<tr><td>序号</td><td colspan="3">工 作 步 骤</td><td colspan="3">注 意 事 项</td></tr>
<tr><td>1</td><td colspan="3">工作前准备</td><td colspan="3"></td></tr>
<tr><td>2</td><td colspan="3">开班前会，布置任务</td><td colspan="3"></td></tr>
<tr><td>3</td><td colspan="3">安全操作运条训练</td><td colspan="3"></td></tr>
<tr><td>4</td><td colspan="3">清理现场</td><td colspan="3"></td></tr>
<tr><td colspan="2" rowspan="3">计划评价</td><td>班级</td><td></td><td>第（ ）组</td><td>组长签字</td><td></td></tr>
<tr><td>教师签字</td><td></td><td>日期</td><td colspan="2"></td></tr>
<tr><td colspan="5">评语：</td></tr>
</table>

4.1.2.3 焊接运条技能训练决策单

<table>
<tr><td>学习场</td><td colspan="3">金 工 实 训</td></tr>
<tr><td>学习情境4</td><td>电气设备接地引下线的焊接</td><td>本环节学时</td><td>6分钟</td></tr>
<tr><td>典型工作过程描述</td><td colspan="3">焊接引弧技能训练—焊接运条技能训练—焊接收尾技能训练—下料—接地引下线搭接平焊</td></tr>
</table>

<table>
<tr><td colspan="6">计 划 对 比</td></tr>
<tr><td>序号</td><td>计划的可行性</td><td>计划的经济性</td><td>计划的可操作性</td><td>计划的实施难度</td><td>综合评价</td></tr>
<tr><td>方案1</td><td></td><td></td><td></td><td></td><td></td></tr>
<tr><td>方案2</td><td></td><td></td><td></td><td></td><td></td></tr>
<tr><td>方案3</td><td></td><td></td><td></td><td></td><td></td></tr>
</table>

<table>
<tr><td rowspan="3">决策评价</td><td>班级</td><td></td><td>第（ ）组</td><td>组长签字</td><td></td></tr>
<tr><td>教师签字</td><td></td><td>日期</td><td colspan="2"></td></tr>
<tr><td colspan="5">评语：</td></tr>
</table>

4.1.2.4 焊接运条技能训练实施单

<table>
<tr><td>学习场</td><td colspan="5">金 工 实 训</td></tr>
<tr><td>学习情境 4</td><td colspan="3">电气设备接地引下线的焊接</td><td>本环节学时</td><td>120 分钟</td></tr>
<tr><td>典型工作过程描述</td><td colspan="5">焊接引弧技能训练—焊接运条技能训练—焊接收尾技能训练—下料—接地引下线搭接平焊</td></tr>
<tr><td>序号</td><td>实施步骤</td><td>工作内容及方法</td><td colspan="2">工艺质量要求</td><td>注意事项</td></tr>
<tr><td>1</td><td>工作前准备</td><td></td><td colspan="2"></td><td></td></tr>
<tr><td>2</td><td>开班前会，布置任务</td><td></td><td colspan="2"></td><td></td></tr>
<tr><td>3</td><td>安全操作运条训练</td><td>1. 焊条沿轴线向熔池方向送进使焊条熔化后，能继续保持电弧的长度不变。
2. 焊条沿焊接方向的纵向移动。
3. 焊条的横向摆动。</td><td colspan="2">要求焊条向熔池方向送进的速度与焊条熔化的速度相等。焊条横向摆动的作用是为获得一定宽度的焊缝，并保证焊缝两侧熔合良好。正常的焊缝宽度一般不超过焊条直径的 2～5 倍。焊条在运行时应该稍作横向摆动，其目的是能获得均匀一致的焊缝成形，同时也是为了控制熔池温度。</td><td>人身触电、电弧伤害、毒气和烟尘伤害、高温伤害。</td></tr>
<tr><td>4</td><td>清理现场</td><td></td><td colspan="2"></td><td></td></tr>
<tr><td colspan="6">实施说明：</td></tr>
</table>

<table>
<tr><td rowspan="3">实施评价</td><td>班级</td><td></td><td>第（ ）组</td><td>组长签字</td><td></td></tr>
<tr><td>教师签字</td><td></td><td>日期</td><td colspan="2"></td></tr>
<tr><td colspan="5">评语：</td></tr>
</table>

4.1.2.5 焊接运条技能训练检查单

学习场	金　工　实　训				
学习情境 4	电气设备接地引下线的焊接			本环节学时	6 分钟
典型工作过程描述	焊接引弧技能训练—焊接运条技能训练—焊接收尾技能训练—下料—接地引下线搭接平焊				
序号	检查项目	检 查 标 准		学生自查	教师检查
1	资讯环节	获得电弧焊接运条安全操作技能的有关信息的来源。			
2	计划环节	制定电弧焊接运条安全操作技能步骤的各个方案。			
3	决策环节	对各种电弧焊接运条安全操作技能的方案衡量对比评分，获取最佳方案。			
4	实施环节	电弧焊接运条安全操作技能训练过程中各个环节按照标准执行，贯彻落实“四不伤害”。			
5	检查环节	各个环节逐一检查，是否有爱岗敬业的主题思想。			
检查评价	班级		第（　　）组	组长签字	
	教师签字		日期		
	评语：				

学习情境 4-13

4.1.3 焊接收尾技能训练

4.1.3.1 焊接收尾技能训练资讯单

学习场	金　工　实　训		
学习情境 4	电气设备接地引下线的焊接	本环节学时	6 分钟
典型工作过程描述	焊接引弧技能训练—焊接运条技能训练—焊接收尾技能训练—下料—接地引下线搭接平焊		
搜集资讯的方式	教师的讲解、互联网查询或者同学间交流		
资讯描述	查看教师提供的资料，获取手工电弧焊接收尾安全操作的技能要求和有关信息，便于安全有效地完成任务。		
对学生的要求	课前充分预习实训作业单，实训过程中穿着实训服装，贯彻落实自己不伤害自己、自己不伤害他人、自己不被他人伤害、保护他人不被伤害的“四不伤害”安全要求，注意严格遵守实训室的各项规章制度。		
参考资料	参考《施工现场临时用电安全技术规范》（JGJ 46—2005）和《电弧焊焊接工艺规程》（GB/T 19867.1—2005）。		

技　能　要　点

接地体（线）的焊接应采用搭接焊，其焊接长度如下：

1. 镀锌扁钢不小于其宽度的（　　），不少于（　　）施焊。当扁钢宽度不同时，搭接长度以（　　）为准。敷设前扁钢需调直，煨弯不得过死，直线段上不应有明显弯曲，并应立放。

2. 镀锌圆钢焊接长度为不小于其直径的（　　）并应（　　）施焊，当直径不同时，搭接长度以（　　　　）的为准。

3. 镀锌圆钢与镀锌扁钢连接时，其长度为不小于圆钢（　　　　　　），必要时采用辅助圆钢或扁钢弯成直角焊接。

4. 镀锌扁钢与镀锌钢管（或角钢）焊接时，除应在其接触部位（　　　）进行焊接外，还应直接将扁钢弯成弧形（或直角形）与钢管（或角钢）焊接，与钢管焊接时，在钢管上的包绕焊接长度不小于扁钢（　　　　　）。

学习情境 4-14

4.1.3.2　焊接收尾技能训练计划单

<table>
<tr><td>学习场</td><td colspan="4">金　工　实　训</td></tr>
<tr><td>学习情境4</td><td colspan="2">电气设备接地引下线的焊接</td><td>本环节学时</td><td>6分钟</td></tr>
<tr><td>典型工作过程描述</td><td colspan="4">焊接引弧技能训练—焊接运条技能训练—焊接收尾技能训练—下料—接地引下线搭接平焊</td></tr>
<tr><td>计划制订的方式</td><td colspan="4">同学之间小组讨论</td></tr>
<tr><td>序号</td><td colspan="2">工 作 步 骤</td><td colspan="2">注 意 事 项</td></tr>
<tr><td>1</td><td colspan="2">工作前准备</td><td colspan="2"></td></tr>
<tr><td>2</td><td colspan="2">开班前会，布置任务</td><td colspan="2"></td></tr>
<tr><td>3</td><td colspan="2">安全操作收尾训练</td><td colspan="2"></td></tr>
<tr><td>4</td><td colspan="2">清理现场</td><td colspan="2"></td></tr>
</table>

<table>
<tr><td rowspan="3">计划评价</td><td>班级</td><td></td><td>第（　　）组</td><td>组长签字</td><td></td></tr>
<tr><td>教师签字</td><td></td><td>日期</td><td colspan="2"></td></tr>
<tr><td colspan="5">评语：</td></tr>
</table>

4.1.3.3 焊接收尾技能训练决策单

<table>
<tr><td>学习场</td><td colspan="5">金　工　实　训</td></tr>
<tr><td>学习情境 4</td><td colspan="3">电气设备接地引下线的焊接</td><td>本环节学时</td><td>6 分钟</td></tr>
<tr><td>典型工作过程描述</td><td colspan="5">焊接引弧技能训练—焊接运条技能训练—焊接收尾技能训练—下料—接地引下线搭接平焊</td></tr>
<tr><td colspan="6">计　划　对　比</td></tr>
<tr><td>序号</td><td>计划的可行性</td><td>计划的经济性</td><td>计划的可操作性</td><td>计划的实施难度</td><td>综合评价</td></tr>
<tr><td>方案 1</td><td></td><td></td><td></td><td></td><td></td></tr>
<tr><td>方案 2</td><td></td><td></td><td></td><td></td><td></td></tr>
<tr><td>方案 3</td><td></td><td></td><td></td><td></td><td></td></tr>
</table>

<table>
<tr><td rowspan="3">决策评价</td><td>班级</td><td></td><td>第（　　）组</td><td>组长签字</td><td></td></tr>
<tr><td>教师签字</td><td></td><td>日期</td><td colspan="2"></td></tr>
<tr><td colspan="5">评语：</td></tr>
</table>

4.1.3.4　焊接收尾技能训练实施单

<table>
<tr><td>学习场</td><td colspan="5">金　工　实　训</td></tr>
<tr><td>学习情境4</td><td colspan="3">电气设备接地引下线的焊接</td><td>本环节学时</td><td>60分钟</td></tr>
<tr><td>典型工作过程描述</td><td colspan="5">焊接引弧技能训练—焊接运条技能训练—焊接收尾技能训练—下料—接地引下线搭接平焊</td></tr>
<tr><td>序号</td><td>实施步骤</td><td>工作内容及方法</td><td colspan="2">工艺质量要求</td><td>注意事项</td></tr>
<tr><td>1</td><td>工作前准备</td><td></td><td colspan="2"></td><td></td></tr>
<tr><td>2</td><td>开班前会，布置任务</td><td></td><td colspan="2"></td><td></td></tr>
<tr><td>3</td><td>安全操作收尾训练</td><td>1. 反复断弧法。
2. 划圈法。
3. 转移收尾法。</td><td colspan="2">1. 反复断弧法：焊条移到焊缝终点时，在弧坑处反复熄弧、引弧数次，直到填满弧坑为止。
2. 划圈法：焊条移到焊缝终点时，在弧坑处作圆圈运动，直到填满弧坑再拉断电弧。
3. 转移法：焊条移到焊缝终点时，在弧坑处稍作停留，将电弧慢慢抬高，再引到焊缝边缘的母材坡口内。</td><td>人身触电、电弧伤害、毒气和烟尘伤害、高温伤害。</td></tr>
<tr><td>4</td><td>清理现场</td><td></td><td colspan="2"></td><td></td></tr>
<tr><td colspan="6">实施说明：</td></tr>
</table>

<table>
<tr><td rowspan="3">实施评价</td><td>班级</td><td></td><td>第（　　）组</td><td>组长签字</td><td></td></tr>
<tr><td>教师签字</td><td></td><td>日　期</td><td colspan="2"></td></tr>
<tr><td colspan="5">评语：</td></tr>
</table>

4.1.3.5 焊接收尾技能训练检查单

<table>
<tr><td colspan="2">学习场</td><td colspan="4">金　工　实　训</td></tr>
<tr><td colspan="2">学习情境 4</td><td colspan="2">电气设备接地引下线的焊接</td><td>本环节学时</td><td>6 分钟</td></tr>
<tr><td colspan="2">典型工作过程描述</td><td colspan="4">焊接引弧技能训练—焊接运条技能训练—焊接收尾技能训练—下料—接地引下线搭接平焊</td></tr>
<tr><td>序号</td><td>检查项目</td><td colspan="2">检 查 标 准</td><td>学生自查</td><td>教师检查</td></tr>
<tr><td>1</td><td>资讯环节</td><td colspan="2">获得电弧焊接收尾安全操作技能的有关信息的来源。</td><td></td><td></td></tr>
<tr><td>2</td><td>计划环节</td><td colspan="2">制定电弧焊接收尾安全操作技能步骤的各个方案。</td><td></td><td></td></tr>
<tr><td>3</td><td>决策环节</td><td colspan="2">对各种电弧焊接收尾安全操作技能的方案衡量对比评分，获取最佳方案。</td><td></td><td></td></tr>
<tr><td>4</td><td>实施环节</td><td colspan="2">电弧焊接收尾安全操作技能训练过程中各个环节按照标准执行，贯彻落实“四不伤害”。</td><td></td><td></td></tr>
<tr><td>5</td><td>检查环节</td><td colspan="2">各个环节逐一检查，是否有爱岗敬业的主题思想。</td><td></td><td></td></tr>
<tr><td colspan="2" rowspan="3">检查评价</td><td>班级</td><td></td><td>第（　　）组</td><td>组长签字</td></tr>
<tr><td>教师签字</td><td></td><td>日期</td><td></td></tr>
<tr><td colspan="4">评语：</td></tr>
</table>

4.1.4　下料

4.1.4.1　下料资讯单

学习场	金　工　实　训		
学习情境4	电气设备接地引下线的焊接	本环节学时	6分钟
典型工作过程描述	焊接引弧技能训练—焊接运条技能训练—焊接收尾技能训练—下料—接地引下线搭接平焊		
搜集资讯的方式	教师的讲解、互联网查询或者同学间交流		
资讯描述	查看教师提供的资料，获取搭接平焊下料的安全操作技能要求和有关信息，便于安全有效地完成任务。		
对学生的要求	课前充分预习实训作业单，实训过程中穿着实训服装，贯彻落实自己不伤害自己、自己不伤害他人、自己不被他人伤害、保护他人不被伤害的“四不伤害”安全要求，注意严格遵守实训室的各项规章制度。		
参考资料	参考《施工现场临时用电安全技术规范》（JGJ 46—2005）和《电弧焊焊接工艺规程》（GB/T 19867.1—2005）。		

图4.1.4中焊接过程有哪些操作是违规的：________________________________

____

____

图4.1.4　现场焊接（1）

4.1.4.2　下料计划单

<table>
<tr><td colspan="2">学习场</td><td colspan="5">金　工　实　训</td></tr>
<tr><td colspan="2">学习情境 4</td><td colspan="3">电气设备接地引下线的焊接</td><td>本环节学时</td><td>6 分钟</td></tr>
<tr><td colspan="2">典型工作过程描述</td><td colspan="5">焊接引弧技能训练—焊接运条技能训练—焊接收尾技能训练—下料—接地引下线搭接平焊</td></tr>
<tr><td colspan="2">计划制订的方式</td><td colspan="5">同学之间小组讨论</td></tr>
<tr><td>序号</td><td>工 作 步 骤</td><td colspan="5">注 意 事 项</td></tr>
<tr><td>1</td><td>工作前准备</td><td colspan="5"></td></tr>
<tr><td>2</td><td>开班前会，布置任务</td><td colspan="5"></td></tr>
<tr><td>3</td><td>下料安全操作</td><td colspan="5"></td></tr>
<tr><td>4</td><td>清理现场</td><td colspan="5"></td></tr>
<tr><td colspan="2" rowspan="3">计划评价</td><td>班级</td><td></td><td>第（　　）组</td><td>组长签字</td><td></td></tr>
<tr><td>教师签字</td><td></td><td>日期</td><td colspan="2"></td></tr>
<tr><td colspan="5">评语：</td></tr>
</table>

4.1.4.3　下料决策单

<table>
<tr><td>学习场</td><td colspan="5">金　工　实　训</td></tr>
<tr><td>学习情境 4</td><td colspan="3">电气设备接地引下线的焊接</td><td>本环节学时</td><td>6 分钟</td></tr>
<tr><td>典型工作过程描述</td><td colspan="5">焊接引弧技能训练—焊接运条技能训练—焊接收尾技能训练—下料—接地引下线搭接平焊</td></tr>
<tr><td colspan="6">计　划　对　比</td></tr>
<tr><td>序号</td><td>计划的可行性</td><td>计划的经济性</td><td>计划的可操作性</td><td>计划的实施难度</td><td>综合评价</td></tr>
<tr><td>方案 1</td><td></td><td></td><td></td><td></td><td></td></tr>
<tr><td>方案 2</td><td></td><td></td><td></td><td></td><td></td></tr>
<tr><td>方案 3</td><td></td><td></td><td></td><td></td><td></td></tr>
</table>

<table>
<tr><td rowspan="3">决策评价</td><td>班级</td><td></td><td>第（　　）组</td><td>组长签字</td><td></td></tr>
<tr><td>教师签字</td><td></td><td>日期</td><td colspan="2"></td></tr>
<tr><td colspan="5">评语：</td></tr>
</table>

学习情境 4-21

4.1.4.4　下料实施单

<table>
<tr><td colspan="2">学习场</td><td colspan="4">金　工　实　训</td></tr>
<tr><td colspan="2">学习情境 4</td><td colspan="2">电气设备接地引下线的焊接</td><td>本环节学时</td><td>40 分钟</td></tr>
<tr><td colspan="2">典型工作过程描述</td><td colspan="4">焊接引弧技能训练—焊接运条技能训练—焊接收尾技能训练—下料—接地引下线搭接平焊</td></tr>
<tr><td>序号</td><td>实施步骤</td><td>工作内容及方法</td><td colspan="2">工艺质量要求</td><td>注意事项</td></tr>
<tr><td>1</td><td>工作前准备</td><td></td><td colspan="2"></td><td></td></tr>
<tr><td>2</td><td>开班前会，
布置任务</td><td></td><td colspan="2"></td><td></td></tr>
<tr><td>3</td><td>下料安全操作</td><td></td><td colspan="2"></td><td></td></tr>
<tr><td>4</td><td>清理现场</td><td></td><td colspan="2"></td><td></td></tr>
<tr><td colspan="6">实施说明：</td></tr>
</table>

<table>
<tr><td rowspan="3">实施评价</td><td>班级</td><td></td><td>第（　　）组</td><td>组长签字</td><td></td></tr>
<tr><td>教师签字</td><td></td><td>日期</td><td colspan="2"></td></tr>
<tr><td colspan="5">评语：</td></tr>
</table>

学习情境 4-22

4.1.4.5　下料检查单

<table>
<tr><td colspan="2">学习场</td><td colspan="4">金　工　实　训</td></tr>
<tr><td colspan="2">学习情境 4</td><td colspan="2">电气设备接地引下线的焊接</td><td>本环节学时</td><td>6 分钟</td></tr>
<tr><td colspan="2">典型工作过程描述</td><td colspan="4">焊接引弧技能训练—焊接运条技能训练—焊接收尾技能训练—下料—接地引下线搭接平焊</td></tr>
<tr><td>序号</td><td>检查项目</td><td colspan="2">检 查 标 准</td><td>学生自查</td><td>教师检查</td></tr>
<tr><td>1</td><td>资讯环节</td><td colspan="2">获得搭焊下料安全操作技能的有关信息的来源。</td><td></td><td></td></tr>
<tr><td>2</td><td>计划环节</td><td colspan="2">制定搭焊下料安全操作技能步骤的各个方案。</td><td></td><td></td></tr>
<tr><td>3</td><td>决策环节</td><td colspan="2">对各种搭焊下料安全操作技能的方案衡量对比评分，获取最佳方案。</td><td></td><td></td></tr>
<tr><td>4</td><td>实施环节</td><td colspan="2">电弧焊接搭焊下料过程中各个环节按照标准执行，贯彻落实“四不伤害”。</td><td></td><td></td></tr>
<tr><td>5</td><td>检查环节</td><td colspan="2">各个环节逐一检查，是否有爱岗敬业的主题思想。</td><td></td><td></td></tr>
<tr><td colspan="2" rowspan="3">检查评价</td><td>班级</td><td></td><td>第（　　）组</td><td>组长签字</td></tr>
<tr><td>教师签字</td><td></td><td>日期</td><td></td></tr>
<tr><td colspan="4">评语：</td></tr>
</table>

4.1.5　接地引下线搭接平焊

4.1.5.1　接地引下线搭接平焊资讯单

学习场	金　工　实　训		
学习情境 4	电气设备接地引下线的焊接	本环节学时	6 分钟
典型工作过程描述	焊接引弧技能训练—焊接运条技能训练—焊接收尾技能训练—下料—接地引下线搭接平焊		
搜集资讯的方式	教师的讲解、互联网查询或者同学间交流		
资讯描述	查看教师提供的资料，获取手工电弧焊搭接安全操作的技能要求和有关信息，便于安全有效地完成任务，参考图 4.1.5。  图 4.1.5　现场焊接（2）		
对学生的要求	课前预习实训作业单，实训过程中穿着实训服装，贯彻落实自己不伤害自己、自己不伤害他人、自己不被他人伤害，保护他人不被伤害的“四不伤害”安全要求，注意严格遵守实训室的各项规章制度。		
参考资料	参考《弧焊设备　第 1 部分：焊接电源》(GB/T 15579.1—2013) 和《电弧焊焊接工艺规程》(GB/T 19867.1—2005)。		

不经历风雨，怎么见彩虹。苦练千百次，操作如有神。

4.1.5.2　接地引下线搭接平焊计划单

<table>
<tr><td>学习场</td><td colspan="6">金　工　实　训</td></tr>
<tr><td>学习情境 4</td><td colspan="4">电气设备接地引下线的焊接</td><td>本环节学时</td><td>6 分钟</td></tr>
<tr><td>典型工作过程描述</td><td colspan="6">焊接引弧技能训练—焊接运条技能训练—焊接收尾技能训练—下料—接地引下线搭接平焊</td></tr>
<tr><td>计划制订的方式</td><td colspan="6">同学之间小组讨论</td></tr>
<tr><td>序号</td><td colspan="2">工 作 步 骤</td><td colspan="4">注 意 事 项</td></tr>
<tr><td>1</td><td colspan="2">工作前准备</td><td colspan="4"></td></tr>
<tr><td>2</td><td colspan="2">开班前会，布置任务</td><td colspan="4"></td></tr>
<tr><td>3</td><td colspan="2">搭焊接安全操作</td><td colspan="4"></td></tr>
<tr><td>4</td><td colspan="2">清理现场</td><td colspan="4"></td></tr>
<tr><td rowspan="3">计划评价</td><td>班级</td><td></td><td>第（　）组</td><td>组长签字</td><td colspan="2"></td></tr>
<tr><td>教师签字</td><td></td><td>日期</td><td colspan="3"></td></tr>
<tr><td colspan="6">评语：</td></tr>
</table>

学习情境 4-25

4.1.5.3 接地引下线搭接平焊决策单

<table>
<tr><td>学习场</td><td colspan="6">金 工 实 训</td></tr>
<tr><td>学习情境4</td><td colspan="3">电气设备接地引下线的焊接</td><td colspan="2">本环节学时</td><td>6分钟</td></tr>
<tr><td>典型工作过程描述</td><td colspan="6">焊接引弧技能训练—焊接运条技能训练—焊接收尾技能训练—下料—接地引下线搭接平焊</td></tr>
<tr><td colspan="7">计 划 对 比</td></tr>
<tr><td>序号</td><td>计划的可行性</td><td>计划的经济性</td><td colspan="2">计划的可操作性</td><td>计划的实施难度</td><td>综合评价</td></tr>
<tr><td>方案1</td><td></td><td></td><td colspan="2"></td><td></td><td></td></tr>
<tr><td>方案2</td><td></td><td></td><td colspan="2"></td><td></td><td></td></tr>
<tr><td>方案3</td><td></td><td></td><td colspan="2"></td><td></td><td></td></tr>
<tr><td rowspan="3">决策评价</td><td>班级</td><td></td><td>第（ ）组</td><td>组长签字</td><td colspan="2"></td></tr>
<tr><td>教师签字</td><td></td><td>日期</td><td colspan="3"></td></tr>
<tr><td colspan="6">评语：</td></tr>
</table>

学习情境 4-26

4.1.5.4　接地引下线搭接平焊实施单

<table>
<tr><td>学习场</td><td colspan="4">金　工　实　训</td></tr>
<tr><td>学习情境 4</td><td colspan="2">电气设备接地引下线的焊接</td><td>本环节学时</td><td>240 分钟</td></tr>
<tr><td>典型工作过程描述</td><td colspan="4">焊接引弧技能训练—焊接运条技能训练—焊接收尾技能训练—下料—接地引下线搭接平焊</td></tr>
</table>

序号	实施步骤	工作内容及方法	工艺质量要求	注意事项
1	工作前准备			
2	开班前会，布置任务			
3	搭焊接安全操作			
4	清理现场	敲掉药皮部分。	使焊接地方呈现出光亮的鱼鳞状。	焊渣伤害。

实施说明：

<table>
<tr><td rowspan="3">实施评价</td><td>班级</td><td></td><td>第（　　）组</td><td>组长签字</td><td></td></tr>
<tr><td>教师签字</td><td></td><td>日期</td><td colspan="2"></td></tr>
<tr><td colspan="5">评语：</td></tr>
</table>

学习情境 4-27

不经历风雨，怎么见彩虹。苦练千百次，操作如有神。

4.1.5.5 接地引下线搭接平焊检查单

<table>
<tr><td colspan="2">学习场</td><td colspan="5">金　工　实　训</td></tr>
<tr><td colspan="2">学习情境 4</td><td colspan="3">电气设备接地引下线的焊接</td><td>本环节学时</td><td>6 分钟</td></tr>
<tr><td colspan="2">典型工作过程描述</td><td colspan="5">焊接引弧技能训练—焊接运条技能训练—焊接收尾技能训练—下料—接地引下线搭接平焊</td></tr>
<tr><td>序号</td><td>检查项目</td><td colspan="3">检 查 标 准</td><td>学生自查</td><td>教师检查</td></tr>
<tr><td>1</td><td>资讯环节</td><td colspan="3">获得电弧焊搭接安全操作技能的有关信息的来源。</td><td></td><td></td></tr>
<tr><td>2</td><td>计划环节</td><td colspan="3">制定电弧焊搭接安全操作技能步骤的各个方案。</td><td></td><td></td></tr>
<tr><td>3</td><td>决策环节</td><td colspan="3">对各种电弧焊搭接安全操作技能的方案衡量对比评分，获取最佳方案。</td><td></td><td></td></tr>
<tr><td>4</td><td>实施环节</td><td colspan="3">电弧焊搭接安全操作技能训练过程中各个环节按照标准执行，贯彻落实“四不伤害”。</td><td></td><td></td></tr>
<tr><td>5</td><td>检查环节</td><td colspan="3">各个环节逐一检查，是否有爱岗敬业的主题思想。</td><td></td><td></td></tr>
<tr><td colspan="2" rowspan="3">检查评价</td><td>班级</td><td></td><td>第（　　）组</td><td>组长签字</td><td></td></tr>
<tr><td>教师签字</td><td></td><td>日期</td><td colspan="2"></td></tr>
<tr><td colspan="5">评语：</td></tr>
</table>

4.2　电气设备接地引下线的焊接学习情境材料工具清单

学习场	金　工　实　训		
学习情境 4	电气设备接地引下线的焊接	本环节学时	660 分钟
典型工作过程描述	焊接引弧技能训练—焊接运条技能训练—焊接收尾技能训练—下料—接地引下线搭接平焊		

序号	名称	作用	数量	型号	使用者

班级		第（　　）组	组长签字	
教师签字		日期		

4.3　电气设备接地引下线的焊接学习情境教学反馈单

<table>
<tr><td>学习场</td><td colspan="4">金　工　实　训</td></tr>
<tr><td>学习情境 4</td><td colspan="2">电气设备接地引下线的焊接</td><td>本环节学时</td><td>660 分钟</td></tr>
<tr><td>典型工作过程描述</td><td colspan="4">焊接引弧技能训练—焊接运条技能训练—焊接收尾技能训练—下料—接地引下线搭接平焊</td></tr>
<tr><td rowspan="6">调查项目</td><td>序号</td><td>调 查 内 容</td><td colspan="2">理 由 描 述</td></tr>
<tr><td></td><td></td><td colspan="2"></td></tr>
<tr><td></td><td></td><td colspan="2"></td></tr>
<tr><td></td><td></td><td colspan="2"></td></tr>
<tr><td></td><td></td><td colspan="2"></td></tr>
<tr><td></td><td></td><td colspan="2"></td></tr>
<tr><td colspan="5">您对本次课程教学的改进意见：</td></tr>
<tr><td>调查信息</td><td>被调查人姓名</td><td></td><td>调查日期</td><td></td></tr>
</table>

4.4　电气设备接地引下线的焊接学习情境分组单

学习场	金 工 实 训					
学习情境 4	电气设备接地引下线的焊接				本环节学时	660 分钟
典型工作过程描述	焊接引弧技能训练—焊接运条技能训练—焊接收尾技能训练—下料—接地引下线搭接平焊					
分组情况	组别	组长	组 员			
分组说明						
班级		教师签字		日期		

学习情境 4-31

4.5 电气设备接地引下线的焊接学习情境评价单

<table>
<tr><td>学习场</td><td colspan="5">金 工 实 训</td></tr>
<tr><td>学习情境 4</td><td colspan="2">电气设备接地引下线的焊接</td><td colspan="2">本环节学时</td><td>20 分钟</td></tr>
<tr><td>典型工作过程描述</td><td colspan="5">焊接引弧技能训练—焊接运条技能训练—焊接收尾技能训练—下料—接地引下线搭接平焊</td></tr>
<tr><td>评价项目</td><td colspan="2">评 价 子 项 目</td><td>学生自评
20%</td><td>组内评价
30%</td><td>教师评价
50%</td></tr>
<tr><td>典型工作环节焊接引弧技能训练</td><td colspan="2">1. 资讯 10 分；2. 计划 10 分；3. 决策 10 分；4. 实施 40 分；5. 检查 10 分；6. 工作态度 20 分。</td><td></td><td></td><td></td></tr>
<tr><td>典型工作环节焊接运条技能训练</td><td colspan="2">1. 资讯 10 分；2. 计划 10 分；3. 决策 10 分；4. 实施 40 分；5. 检查 10 分；6. 工作态度 20 分。</td><td></td><td></td><td></td></tr>
<tr><td>典型工作环节焊接收尾技能训练</td><td colspan="2">1. 资讯 10 分；2. 计划 10 分；3. 决策 10 分；4. 实施 40 分；5. 检查 10 分；6. 工作态度 20 分。</td><td></td><td></td><td></td></tr>
<tr><td>典型工作环节下料</td><td colspan="2">1. 资讯 10 分；2. 计划 10 分；3. 决策 10 分；4. 实施 40 分；5. 检查 10 分；6. 工作态度 20 分。</td><td></td><td></td><td></td></tr>
<tr><td>典型工作环节接地引下线搭接平焊</td><td colspan="2">1. 资讯 10 分；2. 计划 10 分；3. 决策 10 分；4. 实施 40 分；5. 检查 10 分；6. 工作态度 20 分。</td><td></td><td></td><td></td></tr>
<tr><td>最终结果</td><td colspan="2">1. 资讯 10 分；2. 计划 10 分；3. 决策 10 分；4. 实施 40 分；5. 检查 10 分；6. 工作态度 20 分。</td><td></td><td></td><td></td></tr>
<tr><td rowspan="3">评价</td><td>班级</td><td></td><td>第（ ）组</td><td>组长签字</td><td></td></tr>
<tr><td>教师签字</td><td></td><td>日期</td><td colspan="2"></td></tr>
<tr><td colspan="5">评语：</td></tr>
</table>

4.6　教师实施计划单

<table>
<tr><td>学习场</td><td colspan="7">金　工　实　训</td></tr>
<tr><td>学习情境4</td><td colspan="4">电气设备接地引下线的焊接</td><td colspan="2">学时</td><td>660分钟</td></tr>
<tr><td>典型工作过程描述</td><td colspan="7">焊接引弧技能训练—焊接运条技能训练—焊接收尾技能训练—下料—接地引下线搭接平焊</td></tr>
<tr><td>序号</td><td colspan="2">工作与学习步骤</td><td>学时</td><td>使用工具</td><td>地点</td><td>方式</td><td>备注</td></tr>
<tr><td></td><td colspan="2"></td><td></td><td></td><td></td><td></td><td></td></tr>
<tr><td></td><td colspan="2"></td><td></td><td></td><td></td><td></td><td></td></tr>
<tr><td></td><td colspan="2"></td><td></td><td></td><td></td><td></td><td></td></tr>
<tr><td></td><td colspan="2"></td><td></td><td></td><td></td><td></td><td></td></tr>
<tr><td></td><td colspan="2"></td><td></td><td></td><td></td><td></td><td></td></tr>
<tr><td></td><td colspan="2"></td><td></td><td></td><td></td><td></td><td></td></tr>
<tr><td></td><td colspan="2"></td><td></td><td></td><td></td><td></td><td></td></tr>
<tr><td></td><td colspan="2"></td><td></td><td></td><td></td><td></td><td></td></tr>
<tr><td></td><td colspan="2"></td><td></td><td></td><td></td><td></td><td></td></tr>
<tr><td></td><td colspan="2"></td><td></td><td></td><td></td><td></td><td></td></tr>
<tr><td>班级</td><td></td><td>教师签字</td><td colspan="2"></td><td>日期</td><td colspan="2"></td></tr>
</table>

Learning scenario 4
Welding of the electrical equipment earthing down-lead

4.0 Learning worksheet

<table>
<tr><td>Learning field</td><td colspan="6">Metalworking practical training</td></tr>
<tr><td>Learning scenario 4</td><td colspan="3">Welding of the electrical equipment earthing down-lead</td><td colspan="2">Total teaching time</td><td>660 minutes</td></tr>
<tr><td>Description of the typical working processes</td><td colspan="6">Welding arc striking skills training – electrode manipulation skills training – welding finishing off skills training – cutting stock – overlap flat welding of earthing down-lead</td></tr>
<tr><td>Learning objectives</td><td colspan="6">1. Be able to use welding tools correctly.
2. Be able to operate electric welding machines safety and normatively.
3. Master the safe operation skills of welding, including arc striking, electrode manipulation, and finishing off.
4. Master the basic skills of flat welding.</td></tr>
<tr><td>Task description</td><td colspan="3">Overlap two pieces of a flat iron with dimensions of 4mm × 40mm × 200mm and proceed to the manual arc welding of end faces A, B, and C, as indicated in Figure 4.0.1, so as to fulfil the requirements of electrical equipment ground connection.</td><td colspan="3">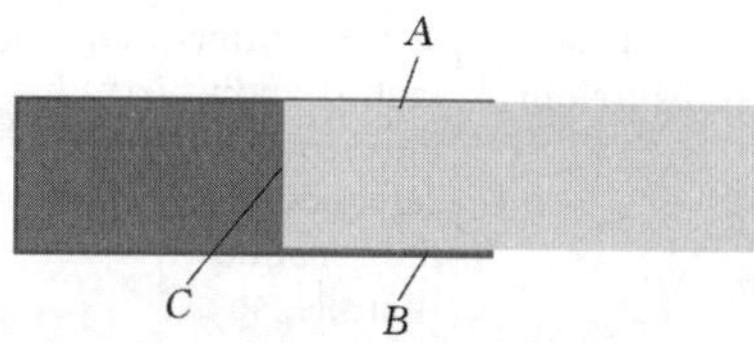
Figure 4.0.1 Welding of three faces</td></tr>
<tr><td>Teaching schedule</td><td>Information: 30 minutes</td><td>Planning: 30 minutes</td><td>Decision:30 minutes</td><td>Implementation: 520 minutes</td><td>Check: 30 minutes</td><td>Assessment: 20 minutes</td></tr>
<tr><td>Requirements for students</td><td colspan="6">Preview the practical training homework sheets before taking the class and use the Internet to find the learning materials related to our practical training. Wear the training uniform throughout the training process. Implement the “4 No-harm Principles” (do not harm yourself, do not harm others, do not get harmed by others, protect others from being harmed) and other safety requirements. Strictly follow the rules and regulations of the training room.</td></tr>
<tr><td>References</td><td colspan="6">Refer to Code for Construction and Acceptance of Grounding Connection Equipment Installation Engineering (GB 50169—2016) and The General Specification for Arc Welding Machines (GB/T 8118—2010).
Three steps of welding   Operation of welding</td></tr>
</table>

4.1 Teaching guidance

Learning field	Metalworking practical training	Learning scenario 4	Welding of the electrical equipment earthing down-lead	Reference frame	*Code for Construction and Acceptance of Grounding Connection Equipment Installation Engineering* (GB 50169—2016)	
General Working processes / Typical working processes	Information	Planning	Decision	Implementation	Check	Assessment
Welding arc striking skills training	Check the *Technical Code for Temporary Electrification on Construction Site* (JGJ 46—2005) to obtain relevant information about the electricity safety code on the construction site, so as to complete the tasks safely and effectively.	Through group discussions among students, prepare an electricity safety code scheme and complete 4.1.1.2 Planning sheet for welding arc striking skills training.	Take integrated evaluation based on the feasibility, economy, operability, and implementation difficulty of the plans and select the best one.	Check relevant information and complete 4.1.1.4 Implementation sheet for welding arc striking skills training.	Conduct self-check by students and check by the teacher according to the check items and check criteria included in 4.1.1.5 Check sheet for welding arc striking skills training.	Conduct self-assessment, peer assessment, and teacher assessment according to the points assigned to the evaluation items, and complete 4.5 Assessment sheet for welding of the electrical equipment earthing down-lead.
Electrode manipulation skills training	Check the *Arc Welding Equipment - Part 1: Welding Power Sources* (GB/T 15579.1—2013) to obtain relevant information about the electricity technical requirements of welding on the construction site, so as to complete the tasks safely and effectively.	Through group discussions among students, prepare an electricity safety code scheme and complete 4.1.2.2 Planning sheet for electrode manipulation skills training.	Take integrated evaluation based on the feasibility, economy, operability, and implementation difficulty of the plans and select the best one.	Check relevant information and complete 4.1.2.4 Implementation sheet for electrode manipulation skills training.	Conduct self-check by students and check by the teacher according to the check items and check criteria included in 4.1.2.5 Check sheet for electrode manipulation skills training	Conduct self-assessment, peer assessment, and teacher assessment according to the points assigned to the evaluation items, and complete 4.5 Assessment sheet for welding of the electrical equipment earthing down-lead.

Continue

Learning field	Metalworking practical training	Learning scenario 4	Welding of the electrical equipment earthing down-lead	Reference frame	*Code for Construction and Acceptance of Grounding Connection Equipment Installation Engineering* (GB 50169—2016)	
General Working processes / Typical working processes	Information	Planning	Decision	Implementation	Check	Assessment
Welding finishing off skills training	Check the *Technical Code for Temporary Electrification on Construction Site* (JGJ 46—2005) to obtain relevant information about the electricity safety code on the construction site, so as to complete the tasks safely and effectively.	Through group discussions among students, prepare an electricity safety code scheme and complete 4.1.3.2 Planning sheet for welding finishing off skills training.	Take integrated evaluation based on the feasibility, economy, operability, and implementation difficulty of the plans and select the best one.	Check relevant information and complete 4.1.3.4 Implementation sheet for welding finishing off skills training.	Conduct self-check by students and check by the teacher according to the check items and check criteria included in 4.1.3.5 Check sheet for welding finishing off skills training.	Conduct self-assessment, peer assessment, and teacher assessment according to the points assigned to the evaluation items, and complete 4.5 Assessment sheet for welding of the electrical equipment earthing down-lead.
Cutting stock	Check the *Technical Code for Temporary Electrification on Construction Site* (JGJ 46—2005) to obtain relevant information about the electricity safety code on the construction site, so as to complete the tasks safely and effectively.	Through group discussions among students, prepare an electricity safety code scheme and complete4.1.4.2 Planning sheet for cutting stock.	Take integrated evaluation based on the feasibility, economy, operability, and implementation difficulty of the plans, and then select the best one.	Check relevant information and complete 4.1.4.4 Implementation sheet for cutting stock.	Conduct self-check by students and check by the teacher according to the check items and check criteria included in 4.1.4.5 Check sheet for cutting stock.	Conduct self-assessment, peer assessment, and teacher assessment according to the points assigned to the evaluation items, and complete 4.5 Assessment sheet for welding of the electrical equipment earthing down-lead.

Continue

Learning field	Metalworking practical training	Learning scenario 4	Welding of the electrical equipment earthing down-lead	Reference frame	*Code for Construction and Acceptance of Grounding Connection Equipment Installation Engineering* (GB 50169—2016)	
General Working processes / Typical working processes	Information	Planning	Decision	Implementation	Check	Assessment
Overlap flat welding of earthing down-lead	Check the *Arc Welding Equipment - Part 1: Welding Power Sources* (GB/T 15579.1—2013) to obtain relevant information about the electricity technical requirements of welding on the construction site, so as to complete the tasks safely and effectively, refer to Figure 4.1.1.	Through group discussions among students, prepare an electricity safety code scheme and complete 4.1.5.2 Planning sheet for overlap flat welding of earthing down lead.	Take integrated evaluation based on the feasibility, economy, operability, and implementation difficulty of the plans, and then select the best one.	Check relevant information and complete 4.1.5.4 Implementation sheet for overlap flat welding of earthing down lead.	Conduct self-check by students and check by the teacher according to the check items and check criteria included in 4.1.5.5 Check sheet for overlap flat welding of earthing down lead.	Conduct self-assessment, peer assessment, and teacher assessment according to the points assigned to the evaluation items, and complete 4.5 Assessment sheet for welding of the electrical equipment earthing down-lead.

Figure 4.1.1　On-site work guidebook

4.1.1 Welding arc striking skills training

4.1.1.1 Information sheet for welding arc striking skills training

<table>
<tr><td>Learning field</td><td colspan="3">Metalworking practical training</td></tr>
<tr><td>Learning scenario 4</td><td>Welding of the electrical equipment earthing down-lead</td><td>Total teaching time</td><td>6 minutes</td></tr>
<tr><td>Description of the typical working processes</td><td colspan="3">Welding arc striking skills training – electrode manipulation skills training – welding finishing off skills training – cutting stock – overlap flat welding of earthing down-lead</td></tr>
<tr><td>Information collection methods</td><td colspan="3">Teacher's explanation, search on the Internet, or communication among students</td></tr>
<tr><td>Information description</td><td colspan="3">Review the materials provided by the instructor to obtain information about the skills and requirements of safe operation of manual arc welding arc striking, so as to complete the tasks safely and effectively, as indicated in Figure 4.1.2.

Figure 4.1.2 Arc striking training</td></tr>
<tr><td>Requirements for students</td><td colspan="3">Preview the practical training homework sheets before taking the class and use the Internet to find the learning materials related to our practical training. Wear the training uniform throughout the training process. Implement the "4 No-harm Principles" (do not harm yourself, do not harm others, do not get harmed by others, protect others from being harmed) and other safety requirements. Strictly follow the rules and regulations of the training room.</td></tr>
<tr><td>References</td><td colspan="3">Refer to Technical Code for Safety of Temporary Electrification on Construction Site (JGJ 46—2005) and Welding Procedure Specification for Arc Welding (GB/T 19867.1—2005).</td></tr>
</table>

4.1.1.2 Planning sheet for welding arc striking skills training

<table>
<tr><td>Learning field</td><td colspan="6">Metalworking practical training</td></tr>
<tr><td>Learning scenario 4</td><td colspan="3">Welding of the electrical equipment earthing down-lead</td><td colspan="2">Total teaching time</td><td>6 minutes</td></tr>
<tr><td>Description of the typical working processes</td><td colspan="6">Welding arc striking skills training – electrode manipulation skills training – welding finishing off skills training – cutting stock – overlap flat welding of earthing down-lead</td></tr>
<tr><td>Planning methods</td><td colspan="6">Group discussion among classmates</td></tr>
<tr><td>Number</td><td colspan="2">Working steps</td><td colspan="4">Notice</td></tr>
<tr><td>1</td><td colspan="2">Preparation before work</td><td colspan="4"></td></tr>
<tr><td>2</td><td colspan="2">Conduct a pre-class meeting, allocate the tasks</td><td colspan="4"></td></tr>
<tr><td>3</td><td colspan="2">Training on safe operation skills of arc striking</td><td colspan="4"></td></tr>
<tr><td>4</td><td colspan="2">Clean up the site</td><td colspan="4"></td></tr>
<tr><td rowspan="3">Plan assessment</td><td>Class</td><td></td><td>Group ( )</td><td>Group leader's signature</td><td colspan="2"></td></tr>
<tr><td>Teacher's signature</td><td></td><td>Date</td><td colspan="3"></td></tr>
<tr><td colspan="6">Comments:</td></tr>
</table>

4.1.1.3 Decision sheet for welding arc striking skills training

<table>
<tr><td>Learning field</td><td colspan="5">Metalworking practical training</td></tr>
<tr><td>Learning scenario 4</td><td colspan="3">Welding of the electrical equipment earthing down-lead</td><td>Total teaching time</td><td>6 minutes</td></tr>
<tr><td>Description of the typical working processes</td><td colspan="5">Welding arc striking skills training – electrode manipulation skills training – welding finishing off skills training – cutting stock – overlap flat welding of earthing down-lead</td></tr>
<tr><td colspan="6">Plan comparison</td></tr>
<tr><td>Number</td><td>Feasibility</td><td>Economy</td><td>Operability</td><td>Implementation difficulty</td><td>Integrated evaluation</td></tr>
<tr><td>Plan 1</td><td></td><td></td><td></td><td></td><td></td></tr>
<tr><td>Plan 2</td><td></td><td></td><td></td><td></td><td></td></tr>
<tr><td>Plan 3</td><td></td><td></td><td></td><td></td><td></td></tr>
</table>

<table>
<tr><td rowspan="3">Decision assessment</td><td>Class</td><td></td><td>Group ( )</td><td>Group leader's signature</td><td></td></tr>
<tr><td>Teacher's signature</td><td></td><td>Date</td><td colspan="2"></td></tr>
<tr><td colspan="5">Comments:</td></tr>
</table>

4.1.1.4 Implementation sheet for welding arc striking skills training

Learning field		Metalworking practical training		
Learning scenario 4		Welding of the electrical equipment earthing down-lead	Total teaching time	60 minutes
Description of the typical working processes		Welding arc striking skills training – electrode manipulation skills training – welding finishing off skills training – cutting stock – overlap flat welding of earthing down-lead		
Number	Implementation steps	Operations and methods	Process quality requirements	Notice
1	Preparation before work			
2	Conduct a pre-class meeting, allocate the tasks			
3	Training on safe operation skills of arc striking	1. Hammering method: Make vertical contact between the welding electrode and weldment. When the endpoint of the welding electrode gently touches the surface of the weldment, quickly lift up the welding electrode, maintain a certain distance, and immediately ignite the arc. 2. Scratching method: First, align the endpoint of the welding electrode to face the weldment, then scratch the weldment surface with the welding electrode. When the arc is ignited, and the metal has not begun to massively meltdown, immediately maintain the distance of 2 – 4mm between the endpoint of the welding electrode and weldment surface, and the arc will be able to combust steadily.	When operating, you must master the time frame and distance of the wrist's up and down movement, ensuring that the arc can combust steadily.	Avoid being harmed by electric shock, arc, poisonous gas, smoke dust, and high temperature.
4	Clean up the site			
Implementation instructions:				

Implementation assessment	Class		Group ( )	Group leader's signature	
	Teacher's signature		Date		
	Comments:				

4.1.1.5 Check sheet for welding arc striking skills training

<table>
<tr><td colspan="2">Learning field</td><td colspan="7">Metalworking practical training</td></tr>
<tr><td colspan="2">Learning scenario 4</td><td colspan="5">Welding of the electrical equipment earthing down-lead</td><td>Total teaching time</td><td>6 minutes</td></tr>
<tr><td colspan="2">Description of the typical working processes</td><td colspan="7">Welding arc striking skills training – electrode manipulation skills training – welding finishing off skills training – cutting stock – overlap flat welding of earthing down-lead</td></tr>
<tr><td>Number</td><td colspan="2">Check items</td><td colspan="3">Check criteria</td><td colspan="2">Self-check by students</td><td>Check by the teacher</td></tr>
<tr><td>1</td><td colspan="2">Information</td><td colspan="3">Obtain the source of information about welding arc striking safe operation skills training.</td><td colspan="2"></td><td></td></tr>
<tr><td>2</td><td colspan="2">Planning</td><td colspan="3">Formulate relevant plans for welding arc striking safe operation skills training.</td><td colspan="2"></td><td></td></tr>
<tr><td>3</td><td colspan="2">Decision</td><td colspan="3">Compare scores of various plans for welding arc striking safe operation skills training and select the best solution.</td><td colspan="2"></td><td></td></tr>
<tr><td>4</td><td colspan="2">Implementation</td><td colspan="3">Operate each step of welding arc striking safe operation skills training in accordance with the standards, and strictly implement the “4 No-harm Principles”.</td><td colspan="2"></td><td></td></tr>
<tr><td>5</td><td colspan="2">Check</td><td colspan="3">Check each step carefully, and check whether the theme idea of dedication to work is followed.</td><td colspan="2"></td><td></td></tr>
<tr><td colspan="2" rowspan="3">Check assessment</td><td colspan="3">Class</td><td colspan="2">Group ( )</td><td>Group leader’s signature</td><td></td></tr>
<tr><td colspan="2">Teacher’s signature</td><td></td><td colspan="2">Date</td><td colspan="2"></td></tr>
<tr><td colspan="7">Comments:</td></tr>
</table>

Without the rain, there would never be rainbows. Practice hard, then achieve remarkably.

4.1.2 Electrode manipulation skills training

4.1.2.1 Information sheet for electrode manipulation skills training

Learning field	Metalworking practical training		
Learning scenario 4	Welding of the electrical equipment earthing down-lead	Total teaching time	6 minutes
Description of the typical working processes	Welding arc striking skills training – electrode manipulation skills training – welding finishing off skills training – cutting stock – overlap flat welding of earthing down-lead		
Information collection methods	Teacher’s explanation, search on the Internet, or communication among students		
Information description	Review the materials provided by the instructor to obtain information about the skills and requirements of safe operation of manual arc electrode welding, so as to complete the tasks safely and effectively, as indicated in Figure 4.1.3.  Figure 4.1.3 Electrode skills training		
Requirements for students	Preview the practical training homework sheets before taking the class and use the Internet to find the learning materials related to our practical training. Wear the training uniform throughout the training process. Implement the “4 No-harm Principles” (do not harm yourself, do not harm others, do not get harmed by others, protect others from being harmed) and other safety requirements. Strictly follow the rules and regulations of the training room.		
References	Refer to *Technical Code for Temporary Electrification on Construction Site* (JGJ 46—2005) and *Welding Procedure Specification for Arc Welding* (GB/T 19867.1—2005).		

4.1.2.2 Planning sheet for electrode manipulation skills training

<table>
<tr><td>Learning field</td><td colspan="5">Metalworking practical training</td></tr>
<tr><td>Learning scenario 4</td><td colspan="3">Welding of the electrical equipment earthing down-lead</td><td>Total teaching time</td><td>6 minutes</td></tr>
<tr><td>Description of the typical working processes</td><td colspan="5">Welding arc striking skills training – electrode manipulation skills training – welding finishing off skills training – cutting stock – overlap flat welding of earthing down-lead</td></tr>
<tr><td>Planning methods</td><td colspan="5">Group discussion among classmates</td></tr>
<tr><td>Number</td><td colspan="2">Working steps</td><td colspan="2">Notice</td><td></td></tr>
<tr><td>1</td><td colspan="2">Preparation before work</td><td colspan="3"></td></tr>
<tr><td>2</td><td colspan="2">Conduct a pre-class meeting, allocate the tasks</td><td colspan="3"></td></tr>
<tr><td>3</td><td colspan="2">Training on safe operation of electrode</td><td colspan="3"></td></tr>
<tr><td>4</td><td colspan="2">Clean up the site</td><td colspan="3"></td></tr>
<tr><td rowspan="3">Plan assessment</td><td>Class</td><td></td><td>Group ( )</td><td>Group leader's signature</td><td></td></tr>
<tr><td>Teacher's signature</td><td></td><td>Date</td><td colspan="2"></td></tr>
<tr><td colspan="5">Comments:</td></tr>
</table>

4.1.2.3 Decision sheet for electrode manipulation skills training

Learning field	Metalworking practical training		
Learning scenario 4	Welding of the electrical equipment earthing down-lead	Total teaching time	6 minutes
Description of the typical working processes	Welding arc striking skills training – electrode manipulation skills training – welding finishing off skills training – cutting stock – overlap flat welding of earthing down-lead		

Plan comparison					
Number	Feasibility	Economy	Operability	Implementation difficulty	Integrated evaluation
Plan 1					
Plan 2					
Plan 3					

Decision assessment	Class		Group ( )	Group leader's signature	
	Teacher's signature		Date		
	Comments:				

Without the rain, there would never be rainbows. Practice hard, then achieve remarkably.

.2.4 Implementation sheet for electrode manipulation skills training

<table>
<tr><td colspan="2">arning field</td><td colspan="4">Metalworking practical training</td></tr>
<tr><td colspan="2">arning scenario 4</td><td colspan="2">Welding of the electrical equipment earthing down-lead</td><td>Total teaching time</td><td>120 minutes</td></tr>
<tr><td colspan="2">scription of the typical rking processes</td><td colspan="4">Welding arc striking skills training – electrode manipulation skills training – welding finishing off skills training – cutting stock – overlap flat welding of earthing down-lead</td></tr>
<tr><td>Number</td><td>Implementation steps</td><td>Operations and methods</td><td colspan="2">Process quality requirements</td><td>Notice</td></tr>
<tr><td>1</td><td>Preparation before work</td><td></td><td colspan="2"></td><td></td></tr>
<tr><td>2</td><td>Conduct a pre-class meeting, allocate the tasks</td><td></td><td colspan="2"></td><td></td></tr>
<tr><td>3</td><td>Training on safe operation skills of electrode manipulation</td><td>1. The welding electrode is fed in along the axis in the direction of the molten pool so that after the welding electrode has melted down, the length of the arc can remain unchanged.
2. The longitudinal movement of the welding electrode along the welding direction.
3. The lateral swing of the welding electrode.</td><td colspan="2">The speed of the welding electrode feeding in the direction of the molten pool should be the same as the melting speed of the welding electrode. The function of the lateral swing of the welding electrode is to obtain a weld of a certain width and ensure both sides of the weld are well fused. The normal width of the weld generally does not exceed 2~5 times the diameter of the welding electrode. The welding electrode should swing slightly laterally during operation, the purpose of which is to obtain a uniform and consistent weld formation, and also to control the temperature of the molten pool.</td><td>Avoid being harmed by electric shock, arc, poisonous gas, smoke dust, and high temperature.</td></tr>
<tr><td>4</td><td>Clean up the site</td><td></td><td colspan="2"></td><td></td></tr>
<tr><td colspan="6">plementation instructions:</td></tr>
<tr><td colspan="2" rowspan="3">plementation assess-nt</td><td>Class</td><td></td><td>Group ( )</td><td>Group leader's signature</td></tr>
<tr><td>Teacher's signature</td><td></td><td>Date</td><td></td></tr>
<tr><td colspan="4">Comments:</td></tr>
</table>

4.1.2.5 Check sheet for electrode manipulation skills training

<table>
<tr><td colspan="2">Learning field</td><td colspan="4">Metalworking practical training</td></tr>
<tr><td colspan="2">Learning scenario 4</td><td colspan="2">Welding of the electrical equipment earthing down-lead</td><td>Total teaching time</td><td>6 minutes</td></tr>
<tr><td colspan="2">Description of the typical working processes</td><td colspan="4">Welding arc striking skills training – electrode manipulation skills training – welding finishing off skills training – cutting stock – overlap flat welding of earthing down-lead</td></tr>
<tr><td>Number</td><td>Check items</td><td colspan="2">Check criteria</td><td>Self-check by students</td><td>Check by the teacher</td></tr>
<tr><td>1</td><td>Information</td><td colspan="2">Obtain the source of information about the safe operation skills of electrode manipulation.</td><td></td><td></td></tr>
<tr><td>2</td><td>Planning</td><td colspan="2">Formulate relevant plans for training on safe operation skills of electrode manipulation.</td><td></td><td></td></tr>
<tr><td>3</td><td>Decision</td><td colspan="2">Compare scores of various plans for training on electrode manipulation skills and select the best solution.</td><td></td><td></td></tr>
<tr><td>4</td><td>Implementation</td><td colspan="2">Operate each step of training on electrode manipulation skills in accordance with the standards, and strictly implement the “4 No-harm Principles”.</td><td></td><td></td></tr>
<tr><td>5</td><td>Check</td><td colspan="2">Check each step carefully, and check whether the theme idea of dedication to work is followed.</td><td></td><td></td></tr>
<tr><td rowspan="3">Check assessment</td><td>Class</td><td></td><td>Group ( )</td><td>Group leader’s signature</td><td></td></tr>
<tr><td>Teacher’s signature</td><td></td><td>Date</td><td colspan="2"></td></tr>
<tr><td colspan="5">Comments:</td></tr>
</table>

4.1.3 Welding finishing off skills training

4.1.3.1 Information sheet for welding finishing off skills training

Learning field	Metalworking practical training		
Learning scenario 4	Welding of the electrical equipment earthing down-lead	Total teaching time	6 minutes
Description of the typical working processes	Welding arc striking skills training – electrode manipulation skills training – welding finishing off skills training – cutting stock – overlap flat welding of earthing down-lead		
Information collection methods	Teacher's explanation, search on the Internet, or communication among students		
Information description	Review the materials provided by the instructor to obtain information about the skills and requirements of safe operation of manual arc welding finishing off, so as to complete the tasks safely and effectively.		
Requirements for students	Preview the practical training homework sheets before taking the class and use the Internet to find the learning materials related to our practical training. Wear the training uniform throughout the training process. Implement the "4 No-harm Principles" (do not harm yourself, do not harm others, do not get harmed by others, protect others from being harmed) and other safety requirements. Strictly follow the rules and regulations of the training room.		
References	Refer to *Technical Code for Temporary Electrification on Construction Site* (JGJ 46—2005) and *Welding Procedure Specification for Arc Welding* (GB/T 19867.1—2005).		

Technical key points

The welding of the earthing electrode (wire) should adopt overlap welding. The length of welding is as follows:

1. The galvanized flat steel is not less than (   ) of its width, not less than (   ) welding's width. When the width of the flat steel is different, the overlap length should be subject to (   ). The flat steel needs to be straightened before laying, the bend should not be too extreme, there should be no obvious bend on the straight section, and it should be placed standing upright.

2. The welding length of the galvanized round steel is not less than (   ) of its diameter and should adopt (   ) welding's width. When the diameter is different, the overlap length should be subject to (   ).

3. When connecting the galvanized flat steel and galvanized round steel, the length is not less than (   ) of the round steel. If necessary, adopt the auxiliary round steel or bend the flat steel to a right angle for welding.

4. When welding the galvanized flat steel and galvanized steel pipe (or angle steel), in addition to welding its contact part (   ), you should also bend the flat steel into an arc shape (or right angle shape) and weld it to the steel pipe (or angle steel). When welding to the steel pipe, the length of wrapper welding on the steel pipe is not less than (   ) of the flat steel.

4.1.3.2 Planning sheet for welding finishing off skills training

<table>
<tr><td>Learning field</td><td colspan="5">Metalworking practical training</td></tr>
<tr><td>Learning scenario 4</td><td colspan="3">Welding of the electrical equipment earthing down-lead</td><td>Total teaching time</td><td>6 minutes</td></tr>
<tr><td>Description of the typical working processes</td><td colspan="5">Welding arc striking skills training – electrode manipulation skills training – welding finishing off skills training – cutting stock – overlap flat welding of earthing down-lead</td></tr>
<tr><td>Planning methods</td><td colspan="5">Group discussion among classmates</td></tr>
<tr><td>Number</td><td colspan="2">Working steps</td><td colspan="3">Notice</td></tr>
<tr><td>1</td><td colspan="2">Preparation before work</td><td colspan="3"></td></tr>
<tr><td>2</td><td colspan="2">Conduct a pre-class meeting, allocate the tasks</td><td colspan="3"></td></tr>
<tr><td>3</td><td colspan="2">Training on safe operation of finishing off</td><td colspan="3"></td></tr>
<tr><td>4</td><td colspan="2">Clean up the site</td><td colspan="3"></td></tr>
<tr><td rowspan="3">Plan assessment</td><td>Class</td><td></td><td>Group ( )</td><td>Group leader's signature</td><td></td></tr>
<tr><td>Teacher's signature</td><td></td><td>Date</td><td colspan="2"></td></tr>
<tr><td colspan="5">Comments:</td></tr>
</table>

4.1.3.3 Decision sheet for welding finishing off skills training

<table>
<tr><td>Learning field</td><td colspan="5">Metalworking practical training</td></tr>
<tr><td>Learning scenario 4</td><td colspan="3">Welding of the electrical equipment earthing down-lead</td><td>Total teaching time</td><td>6 minutes</td></tr>
<tr><td>Description of the typical working processes</td><td colspan="5">Welding arc striking skills training – electrode manipulation skills training – welding finishing off skills training – cutting stock – overlap flat welding of earthing down-lead</td></tr>
<tr><td colspan="6">Plan comparison</td></tr>
<tr><td>Number</td><td>Feasibility</td><td>Economy</td><td>Operability</td><td>Implementation difficulty</td><td>Integrated evaluation</td></tr>
<tr><td>Plan 1</td><td></td><td></td><td></td><td></td><td></td></tr>
<tr><td>Plan 2</td><td></td><td></td><td></td><td></td><td></td></tr>
<tr><td>Plan 3</td><td></td><td></td><td></td><td></td><td></td></tr>
<tr><td rowspan="3">Decision assessment</td><td>Class</td><td></td><td>Group ( )</td><td>Group leader's signature</td><td></td></tr>
<tr><td>Teacher's signature</td><td></td><td>Date</td><td colspan="2"></td></tr>
<tr><td colspan="5">Comments:</td></tr>
</table>

4.1.3.4 Implementation sheet for welding finishing off skills training

Learning field	Metalworking practical training		
Learning scenario 4	Welding of the electrical equipment earthing down-lead	Total teaching time	60 minutes
Description of the typical working processes	Welding arc striking skills training – electrode manipulation skills training – welding finishing off skills training – cutting stock – overlap flat welding of earthing down-lead		

Number	Implementation steps	Operations and methods	Process quality requirements	Notice
1	Preparation before work			
2	Conduct a pre-class meeting, allocate the tasks			
3	Training on safe operation skills of finishing off	1. Repeated arc breaking method. 2. Circling method. 3. Shift finishing off method.	1. Repeated arc breaking method: When the welding electrode is moved to the end of the weld, quench, and ignite the arc repeatedly at the arc pit several times until the arc pit is filled. 2. Circling method: When the welding electrode is moved to the end of the weld, perform a circular motion at the arc pit until the arc pit is filled, then pull off the arc. 3. Shift method: When the welding electrode is moved to the end of the weld, stop for a while at the arc pit, slowly raise the arc, and then lead into the parent metal groove at the edge of the weld.	Avoid being harmed by electric shock, arc, poisonous gas, smoke dust, and high temperature.
4	Clean up the site			

Implementation instructions:

Implementation assessment	Class		Group ( )	Group leader's signature	
	Teacher's signature		Date		
	Comments:				

4.1.3.5 Check sheet for welding finishing off skills training

<table>
<tr><td colspan="2">Learning field</td><td colspan="4">Metalworking practical training</td></tr>
<tr><td colspan="2">Learning scenario 4</td><td colspan="2">Welding of the electrical equipment earthing down-lead</td><td>Total teaching time</td><td>6 minutes</td></tr>
<tr><td colspan="2">Description of the typical working processes</td><td colspan="4">Welding arc striking skills training – electrode manipulation skills training – welding finishing off skills training – cutting stock – overlap flat welding of earthing down-lead</td></tr>
<tr><td>Number</td><td>Check items</td><td colspan="2">Check criteria</td><td>Self-check by students</td><td>Check by the teacher</td></tr>
<tr><td>1</td><td>Information</td><td colspan="2">Obtain the source of information about arc welding finishing off safe operation skills training.</td><td></td><td></td></tr>
<tr><td>2</td><td>Planning</td><td colspan="2">Formulate relevant plans for arc welding finishing off safe operation skills training.</td><td></td><td></td></tr>
<tr><td>3</td><td>Decision</td><td colspan="2">Compare scores of various plans for arc welding finishing off safe operation skills training and select the best solution.</td><td></td><td></td></tr>
<tr><td>4</td><td>Implementation</td><td colspan="2">Operate each step of arc welding finishing off safe operation skills training in accordance with the standards, and strictly implement the “4 No-harm Principles”.</td><td></td><td></td></tr>
<tr><td>5</td><td>Check</td><td colspan="2">Check each step carefully, and check whether the theme idea of dedication to work is followed.</td><td></td><td></td></tr>
<tr><td rowspan="3">Check assessment</td><td>Class</td><td></td><td>Group ( )</td><td>Group leader’s signature</td><td></td></tr>
<tr><td>Teacher’s signature</td><td></td><td>Date</td><td></td><td></td></tr>
<tr><td colspan="5">Comments:</td></tr>
</table>

4.1.4 Cutting stock

4.1.4.1 Information sheet for cutting stock

Learning field	Metalworking practical training		
Learning scenario 4	Welding of the electrical equipment earthing down-lead	Total teaching time	6 minutes
Description of the typical working processes	Welding arc striking skills training – electrode manipulation skills training – welding finishing off skills training – cutting stock – overlap flat welding of earthing down-lead		
Information collection methods	Teacher's explanation, search on the Internet, or communication among students		
Information description	Review the materials provided by the instructor to obtain information about the skills and requirements of cutting stock of overlap flat welding, so as to complete the tasks safely and effectively.		
Requirements for students	Preview the practical training homework sheets before taking the class and use the Internet to find the learning materials related to our practical training. Wear the training uniform throughout the training process. Implement the "4 No-harm Principles" (do not harm yourself, do not harm others, do not get harmed by others, protect others from being harmed) and other safety requirements. Strictly follow the rules and regulations of the training room.		
References	Refer to *Technical Code for Temporary Electrification on Construction Site* (JGJ 46—2005) and *Welding Procedure Specification for Arc Welding* (GB/T 19867.1—2005).		

Figure 4.1.4 shows the welding process, please point out any violations: ____________________

____

____

Figure 4.1.4 On-site welding (1)

4.1.4.2 Planning sheet for cutting stock

<table>
<tr><td colspan="2">Learning field</td><td colspan="5">Metalworking practical training</td></tr>
<tr><td colspan="2">Learning scenario 4</td><td colspan="3">Welding of the electrical equipment earthing down-lead</td><td>Total teaching time</td><td>6 minutes</td></tr>
<tr><td colspan="2">Description of the typical working processes</td><td colspan="5">Welding arc striking skills training – electrode manipulation skills training – welding finishing off skills training – cutting stock – overlap flat welding of earthing down-lead</td></tr>
<tr><td colspan="2">Planning methods</td><td colspan="5">Group discussion among classmates</td></tr>
<tr><td>Number</td><td colspan="3">Working steps</td><td colspan="3">Notice</td></tr>
<tr><td>1</td><td colspan="3">Preparation before work</td><td colspan="3"></td></tr>
<tr><td>2</td><td colspan="3">Conduct a pre-class meeting, allocate the tasks</td><td colspan="3"></td></tr>
<tr><td>3</td><td colspan="3">Safe operation of cutting stock</td><td colspan="3"></td></tr>
<tr><td>4</td><td colspan="3">Clean up the site</td><td colspan="3"></td></tr>
<tr><td colspan="2" rowspan="3">Plan assessment</td><td>Class</td><td></td><td>Group ( )</td><td>Group leader’s signature</td><td></td></tr>
<tr><td>Teacher’s signature</td><td></td><td>Date</td><td></td><td></td></tr>
<tr><td colspan="5">Comments:</td></tr>
</table>

4.1.4.3 Decision sheet for cutting stock

<table>
<tr><td>Learning field</td><td colspan="5">Metalworking practical training</td></tr>
<tr><td>Learning scenario 4</td><td colspan="3">Welding of the electrical equipment earthing down-lead</td><td>Total teaching time</td><td>6 minutes</td></tr>
<tr><td>Description of the typical working processes</td><td colspan="5">Welding arc striking skills training – electrode manipulation skills training – welding finishing off skills training – cutting stock – overlap flat welding of earthing down-lead</td></tr>
<tr><td colspan="6">Plan comparison</td></tr>
<tr><td>Number</td><td>Feasibility</td><td>Economy</td><td>Operability</td><td>Implementation difficulty</td><td>Integrated evaluation</td></tr>
<tr><td>Plan 1</td><td></td><td></td><td></td><td></td><td></td></tr>
<tr><td>Plan 2</td><td></td><td></td><td></td><td></td><td></td></tr>
<tr><td>Plan 3</td><td></td><td></td><td></td><td></td><td></td></tr>
<tr><td rowspan="3">Decision assessment</td><td>Class</td><td></td><td>Group ( )</td><td>Group leader's signature</td><td></td></tr>
<tr><td>Teacher's signature</td><td></td><td>Date</td><td colspan="2"></td></tr>
<tr><td colspan="5">Comments:</td></tr>
</table>

4.1.4.4 Implementation sheet for cutting stock

<table>
<tr><td colspan="2">Learning field</td><td colspan="4">Metalworking practical training</td></tr>
<tr><td colspan="2">Learning scenario 4</td><td colspan="2">Welding of the electrical equipment earthing down-lead</td><td>Total teaching time</td><td>40 minutes</td></tr>
<tr><td colspan="2">Description of the typical working processes</td><td colspan="4">Welding arc striking skills training – electrode manipulation skills training – welding finishing off skills training – cutting stock – overlap flat welding of earthing down-lead</td></tr>
<tr><td>Number</td><td>Implementation steps</td><td>Operations and methods</td><td>Process quality requirements</td><td colspan="2">Notice</td></tr>
<tr><td>1</td><td>Preparation before work</td><td></td><td></td><td colspan="2"></td></tr>
<tr><td>2</td><td>Conduct a pre-class meeting, allocate the tasks</td><td></td><td></td><td colspan="2"></td></tr>
<tr><td>3</td><td>Safe operation of cutting stock</td><td></td><td></td><td colspan="2"></td></tr>
<tr><td>4</td><td>Clean up the site</td><td></td><td></td><td colspan="2"></td></tr>
<tr><td colspan="6">Implementation instructions:</td></tr>
<tr><td rowspan="3">Implementation assessment</td><td>Class</td><td></td><td>Group ( )</td><td>Group leader's signature</td><td></td></tr>
<tr><td>Teacher's signature</td><td></td><td>Date</td><td></td><td></td></tr>
<tr><td colspan="5">Comments:</td></tr>
</table>

Without the rain, there would never be rainbows. Practice hard, then achieve remarkably.

4.1.4.5　Check sheet for cutting stock

<table>
<tr><td>Learning field</td><td colspan="6">Metalworking practical training</td></tr>
<tr><td>Learning scenario 4</td><td colspan="3">Welding of the electrical equipment earthing down-lead</td><td>Total teaching time</td><td colspan="2">6 minutes</td></tr>
<tr><td>Description of the typical working processes</td><td colspan="6">Welding arc striking skills training – electrode manipulation skills training – welding finishing off skills training – cutting stock – overlap flat welding of earthing down-lead</td></tr>
<tr><td>Number</td><td colspan="2">Check items</td><td>Check criteria</td><td colspan="2">Self-check by students</td><td>Check by the teacher</td></tr>
<tr><td>1</td><td colspan="2">Information</td><td>Obtain the source of information about safe operation skills training of cutting stock of overlap welding.</td><td colspan="2"></td><td></td></tr>
<tr><td>2</td><td colspan="2">Planning</td><td>Formulate relevant plans for safe operation skills training of cutting stock of overlap welding.</td><td colspan="2"></td><td></td></tr>
<tr><td>3</td><td colspan="2">Decision</td><td>Compare scores of various plans for safe operation skills training of cutting stock of overlap welding and select the best solution.</td><td colspan="2"></td><td></td></tr>
<tr><td>4</td><td colspan="2">Implementation</td><td>Operate each step of the process of cutting stock of overlap welding in accordance with the standards, and strictly implement the “4 No-harm Principles”.</td><td colspan="2"></td><td></td></tr>
<tr><td>5</td><td colspan="2">Check</td><td>Check each step carefully, and check whether the theme idea of dedication to work is followed.</td><td colspan="2"></td><td></td></tr>
<tr><td colspan="2" rowspan="3">Check assessment</td><td>Class</td><td></td><td>Group ( )</td><td>Group leader’s signature</td><td></td></tr>
<tr><td>Teacher’s signature</td><td></td><td>Date</td><td></td><td></td></tr>
<tr><td colspan="5">Comments:</td></tr>
</table>

4.1.5 Overlap flat welding of earthing down-lead

4.1.5.1 Information sheet for overlap flat welding of earthing down-lead

Learning field	Metalworking practical training		
Learning scenario 4	Welding of the electrical equipment earthing down-lead	Total teaching time	6 minutes
Description of the typical working processes	Welding arc striking skills training – electrode manipulation skills training – welding finishing off skills training – cutting stock – overlap flat welding of earthing down-lead		
Information collection methods	Teacher's explanation, search on the Internet, or communication among students		
Information description	Review the materials provided by the instructor to obtain information about the skills and requirements of safe operation of manual arc overlap welding, so as to complete the tasks safely and effectively. You can refer to Figure 4.1.5.  Figure 4.1.5 On-site welding (2)		
Requirements for students	Preview the practical training homework sheets before taking the class and use the Internet to find the learning materials related to our practical training. Wear the training uniform throughout the training process. Implement the "4 No-harm Principles" (do not harm yourself, do not harm others, do not get harmed by others, protect others from being harmed) and other safety requirements. Strictly follow the rules and regulations of the training room.		
References	Refer to *Arc Welding Equipment — Part 1: Welding Power Sources* (GB/T 15579.1—2013) and *Welding Procedure Specification for Arc Welding* (GB/T 19867.1—2005).		

4.1.5.2 Planning sheet for overlap flat welding of earthing down-lead

<table>
<tr><td>Learning field</td><td colspan="5">Metalworking practical training</td></tr>
<tr><td>Learning scenario 4</td><td colspan="2">Welding of the electrical equipment earthing down-lead</td><td colspan="2">Total teaching time</td><td>6 minutes</td></tr>
<tr><td>Description of the typical working processes</td><td colspan="5">Welding arc striking skills training – electrode manipulation skills training – welding finishing off skills training – cutting stock – overlap flat welding of earthing down-lead</td></tr>
<tr><td>Planning methods</td><td colspan="5">Group discussion among classmates</td></tr>
<tr><td>Number</td><td colspan="2">Working steps</td><td colspan="3">Notice</td></tr>
<tr><td>1</td><td colspan="2">Preparation before work</td><td colspan="3"></td></tr>
<tr><td>2</td><td colspan="2">Conduct a pre-class meeting, allocate the tasks</td><td colspan="3"></td></tr>
<tr><td>3</td><td colspan="2">Safe operation of overlap welding</td><td colspan="3"></td></tr>
<tr><td>4</td><td colspan="2">Clean up the site</td><td colspan="3"></td></tr>
<tr><td rowspan="3">Plan assessment</td><td>Class</td><td></td><td>Group ( )</td><td>Group leader's signature</td><td></td></tr>
<tr><td>Teacher's signature</td><td></td><td>Date</td><td></td><td></td></tr>
<tr><td colspan="5">Comments:</td></tr>
</table>

4.1.5.3 Decision sheet for overlap flat welding of earthing down-lead

<table>
<tr><td>Learning field</td><td colspan="6">Metalworking practical training</td></tr>
<tr><td>Learning scenario 4</td><td colspan="3">Welding of the electrical equipment earthing down-lead</td><td colspan="2">Total teaching time</td><td>6 minutes</td></tr>
<tr><td>Description of the typical working processes</td><td colspan="6">Welding arc striking skills training – electrode manipulation skills training – welding finishing off skills training – cutting stock – overlap flat welding of earthing down-lead</td></tr>
<tr><td colspan="7">Plan comparison</td></tr>
<tr><td>Number</td><td>Feasibility</td><td>Economy</td><td>Operability</td><td colspan="2">Implementation difficulty</td><td>Integrated evaluation</td></tr>
<tr><td>Plan 1</td><td></td><td></td><td></td><td colspan="2"></td><td></td></tr>
<tr><td>Plan 2</td><td></td><td></td><td></td><td colspan="2"></td><td></td></tr>
<tr><td>Plan 3</td><td></td><td></td><td></td><td colspan="2"></td><td></td></tr>
<tr><td rowspan="3">Decision assessment</td><td>Class</td><td></td><td>Group ( )</td><td>Group leader's signature</td><td colspan="2"></td></tr>
<tr><td>Teacher's signature</td><td></td><td>Date</td><td colspan="3"></td></tr>
<tr><td colspan="6">Comments:</td></tr>
</table>

4.1.5.4 Implementation sheet for overlap flat welding of earthing down-lead

Learning field	Metalworking practical training		
Learning scenario 4	Welding of the electrical equipment earthing down-lead	Total teaching time	240 minutes
Description of the typical working processes	Welding arc striking skills training – electrode manipulation skills training – welding finishing off skills training – cutting stock – overlap flat welding of earthing down-lead		

Number	Implementation steps	Operations and methods	Process quality requirements	Notice
1	Preparation before work			
2	Conduct a pre-class meeting, allocate the tasks			
3	Safe operation of overlap welding			
4	Clean up the site	Knock off the coating part.	Gives a shiny, fish-scaled appearance to the weld area.	Avoid being harmed by welding slag.

Implementation instructions:

Imple-mentation assessment	Class		Group ( )	Group leader's signature	
	Teacher's signature		Date		
	Comments:				

4.1.5.5 Check sheet for overlap flat welding of earthing downlead

<table>
<tr><td>Learning field</td><td colspan="5">Metalworking practical training</td></tr>
<tr><td>Learning scenario 4</td><td colspan="3">Welding of the electrical equipment earthing down-lead</td><td>Total teaching time</td><td>6 minutes</td></tr>
<tr><td>Description of the typical working processes</td><td colspan="5">Welding arc striking skills training – electrode manipulation skills training – welding finishing off skills training – cutting stock – overlap flat welding of earthing down-lead</td></tr>
<tr><td>Number</td><td>Check items</td><td>Check criteria</td><td colspan="2">Self-check by students</td><td>Check by the teacher</td></tr>
<tr><td>1</td><td>Information</td><td>Obtain the source of information about arc overlap welding safe operation skills training.</td><td colspan="2"></td><td></td></tr>
<tr><td>2</td><td>Planning</td><td>Formulate relevant plans for arc overlap welding safe operation skills training.</td><td colspan="2"></td><td></td></tr>
<tr><td>3</td><td>Decision</td><td>Compare scores of various plans for arc overlap welding safe operation skills training and select the best solution.</td><td colspan="2"></td><td></td></tr>
<tr><td>4</td><td>Implementation</td><td>Operate each step of arc overlap welding safe operation skills training in accordance with the standards, and strictly implement the “4 No-harm Principles”.</td><td colspan="2"></td><td></td></tr>
<tr><td>5</td><td>Check</td><td>Check each step carefully, and check whether the theme idea of dedication to work is followed.</td><td colspan="2"></td><td></td></tr>
<tr><td rowspan="3">Check assessment</td><td>Class</td><td></td><td>Group ( )</td><td>Group leader’s signature</td><td></td></tr>
<tr><td>Teacher’s signature</td><td></td><td>Date</td><td colspan="2"></td></tr>
<tr><td colspan="5">Comments:</td></tr>
</table>

4.2 List of materials and tools for welding of the electrical equipment earthing down-lead

Learning field	Metalworking practical training				
Learning scenario 4	Welding of the electrical equipment earthing down-lead		Total teaching time	660 minutes	
Description of the typical working processes	Welding arc striking skills training – electrode manipulation skills training – welding finishing off skills training – cutting stock – overlap flat welding of earthing down-lead				
Number	Name	Function	Quantity	Model	User
Class		Group ( )	Group leader's signature		
Teacher's signature		Date			

4.3 Teaching feedback sheet for welding of the electrical equipment earthing down-lead

<table>
<tr><td>Learning field</td><td colspan="4">Metalworking practical training</td></tr>
<tr><td>Learning scenario 4</td><td colspan="2">Welding of the electrical equipment earthing down-lead</td><td>Total teaching time</td><td>660 minutes</td></tr>
<tr><td>Description of the typical working processes</td><td colspan="4">Welding arc striking skills training – electrode manipulation skills training – welding finishing off skills training – cutting stock – overlap flat welding of earthing down-lead</td></tr>
<tr><td rowspan="6">Survey items</td><td>Number</td><td colspan="2">Survey contents</td><td>Description of the reasons</td></tr>
<tr><td></td><td colspan="2"></td><td></td></tr>
<tr><td></td><td colspan="2"></td><td></td></tr>
<tr><td></td><td colspan="2"></td><td></td></tr>
<tr><td></td><td colspan="2"></td><td></td></tr>
<tr><td></td><td colspan="2"></td><td></td></tr>
<tr><td colspan="5">Your feedback on how to improve the teaching of this course:</td></tr>
<tr><td>Survey information</td><td>Respondent's name</td><td></td><td>Survey date</td><td></td></tr>
</table>

4.4 Grouping sheet for welding of the electrical equipment earthing down-lead

Learning field	Metalworking practical training					
Learning scenario 4	Welding of the electrical equipment earthing down-lead			Total teaching time	660 minutes	
Description of the typical working processes	Welding arc striking skills training – electrode manipulation skills training – welding finishing off skills training – cutting stock – overlap flat welding of earthing down-lead					
Grouping	Group	Group leader	Group members			
Grouping instructions						
Class		Teacher's signature		Date		

Without the rain, there would never be rainbows. Practice hard, then achieve remarkably.

4.5 Assessment sheet for welding of the electrical equipment earthing down-lead

Learning field	Metalworking practical training			
Learning scenario 4	Welding of the electrical equipment earthing down-lead	Total teaching time	20 minutes	
Description of the typical working processes	Welding arc striking skills training – electrode manipulation skills training – welding finishing off skills training – cutting stock – overlap flat welding of earthing down-lead			
Assessment items	Assessment subitems	Self-assessment by students 20%	Intra-group assessment 30%	Assessment by the teacher 50%
Welding arc striking skills training	1. Information: 10 points; 2. Planning: 10 points; 3. Decision: 10 points; 4. Implementation: 40 points; 5. Check: 10 points; 6. Work attitude: 20 points.			
Electrode manipulation skills training	1. Information: 10 points; 2. Planning: 10 points; 3. Decision: 10 points; 4. Implementation: 40 points; 5. Check: 10 points; 6. Work attitude: 20 points.			
Welding finishing off skills training	1. Information: 10 points; 2. Planning: 10 points; 3. Decision: 10 points; 4. Implementation: 40 points; 5. Check: 10 points; 6. Work attitude: 20 points.			
Cutting stock	1. Information: 10 points; 2. Planning: 10 points; 3. Decision: 10 points; 4. Implementation: 40 points; 5. Check: 10 points; 6. Work attitude: 20 points.			
Overlap flat welding of earthing down-lead	1. Information: 10 points; 2. Planning: 10 points; 3. Decision: 10 points; 4. Implementation: 40 points; 5. Check: 10 points; 6. Work attitude: 20 points.			

Continue

<table>
<tr><th>Assessment items</th><th colspan="2">Assessment subitems</th><th>Self-assessment by students 20%</th><th colspan="2">Intra-group assessment 30%</th><th>Assessment by the teacher 50%</th></tr>
<tr><td>Final results</td><td colspan="2">1. Information: 10 points;
2. Planning: 10 points;
3. Decision: 10 points;
4. Implementation: 40 points;
5. Check: 10 points;
6. Work attitude: 20 points.</td><td></td><td colspan="2"></td><td></td></tr>
<tr><td rowspan="3">Assessment</td><td>Class</td><td></td><td>Group ( )</td><td>Group leader's signature</td><td colspan="2"></td></tr>
<tr><td>Teacher's signature</td><td></td><td>Date</td><td colspan="3"></td></tr>
<tr><td colspan="6">Comments:</td></tr>
</table>

4.6 Teacher's implementation planning sheet

Learning field	Metalworking practical training					
Learning scenario 4	Welding of the electrical equipment earthing down-lead			Teaching time	660 minutes	
Description of the typical working processes	Welding arc striking skills training – electrode manipulation skills training – welding finishing off skills training – cutting stock – overlap flat welding of earthing down-lead					
Number	Working and learning steps	Teaching time	Tools	Locations	Methods	Notes
Class		Teacher's signature		Date		

参 考 文 献

[1] 姜大源. 职业教育学新论 [M]. 北京：教育科学出版社，2007.

[2] 姜大源. 工作过程导向的高职课程开发探索与实践——国家示范性高等职业院校课程开发案例汇编 [M]. 北京：高等教育出版社，2008.

[3] 戴士宏. 职业教育课程教学改革 [M]. 北京：清华大学出版社，2007.

[4] 中华人民共和国国家质量监督检验检疫总局. GB 26860—2011 电力安全工作规程　发电厂和变电站电气部分 [S]. 北京：中国标准出版社，2011.

[5] 中华人民共和国建设部. JGJ 46—2005 施工现场临时用电安全技术规范 [S]. 北京：中国建筑工业出版社，2005.

[6] 中华人民共和国国家质量监督检验检疫总局. GB/T 15579.1—2013 弧焊设备　第 1 部分：焊接电源 [S]. 北京：中国质检出版社，2013.

[7] 中华人民共和国国家质量监督检验检疫总局. GB/T 19867.1—2005 电弧焊焊接工艺规程 [S]. 北京：中国质检出版社，2005.

[8] 中华人民共和国住房和城乡建设部. GB 50169—2016 电气装置安装工程　接地装置施工及验收规范 [S]. 北京：中国计划出版社，2016.

[9] 中国国家标准化管理委员会. GB/T 8118—2010 电弧焊机通用技术条件 [S]. 北京：中国质检出版社，2010.

[10] 中华人民共和国国家质量监督检验检疫总局. GB/T 6170—2015 1 型六角螺母 [S]. 北京：中国标准出版社，2015.

References

[1] 姜大源 . 职业教育学新论 [M]. 北京：教育科学出版社，2007.

[2] 姜大源 . 工作过程导向的高职课程开发探索与实践——国家示范性高等职业院校课程开发案例汇编 [M]. 北京：高等教育出版社，2008.

[3] 戴士宏 . 职业教育课程教学改革 [M]. 北京：清华大学出版社，2007.

[4] 中华人民共和国国家质量监督检验检疫总局 . GB 26860—2011 电力安全工作规程　发电厂和变电站电气部分 [S]. 北京：中国标准出版社，2011.

[5] 中华人民共和国建设部 . JG J46—2005 施工现场临时用电安全技术规范 [S]. 北京：中国建筑工业出版社，2005.

[6] 中华人民共和国国家质量监督检验检疫总局 . GB/T 15579.1—2013 弧焊设备　第 1 部分：焊接电源 [S]. 北京：中国质检出版社，2013.

[7] 中华人民共和国国家质量监督检验检疫总局 . GB/T 19867.1—2005 电弧焊焊接工艺规程 [S]. 北京：中国质检出版社，2005.

[8] 中华人民共和国住房和城乡建设部 . GB 50169—2016 实电气装置安装工程　接地装置施工及验收规范 [S]. 北京：中国计划出版社，2016.

[9] 中国国家标准化管理委员会 . GB/T 8118—2010 电弧焊机通用技术条件 [S]. 北京：中国质检出版社，2010.

[10] 中华人民共和国国家质量监督检验检疫总局 . GB/T 6170—2015 1 型六角螺母 [S]. 北京：中国标准出版社，2015.